3D PRINTING

BUILD YOUR OWN 3D PRINTER AND PRINT YOUR OWN 3D OBJECTS

James Floyd Kelly

800 East 96th Street,
Indianapolis, Indiana 46240 USA

3D Printing: Build Your Own 3D Printer and Print Your Own 3D Objects

ISBN-13: 978-0-7897-5235-2
ISBN-10: 0-7897-5235-2

Library of Congress Control Number: 2013949972

Printed in the United States of America

First printing October 2013

Bulk Sales

Que Publishing offers excellent discounts on this book when ordered in quantity for bulk purchases or special sales. For more information, please contact

U.S. Corporate and Government Sales
1-800-382-3419
corpsales@pearsontechgroup.com

For sales outside of the U.S., please contact

International Sales
international@pearsoned.com

Editor-in-Chief
Greg Wiegand

Executive Editor
Rick Kughen

Development Editors
William Abner
Todd Brakke

Managing Editor
Sandra Schroeder

Project Editor
Mandie Frank

Copy Editor
Barbara Hacha

Indexer
Lisa Stumpf

Proofreader
Dan Knott

Technical Editor
John Ray

Publishing Coordinator
Kristen Watterson

Designer
Mark Shirar

Compositor
Mary Sudul

Contents at a Glance

Table of Contents

About the Author

James Floyd Kelly is a technology writer with degrees in English and Industrial Engineering. James has written on a wide variety of topics, including programming for kids, LEGO robotics, open-source software, and building 3D printers. James is a DIYer—a tinkerer and a maker who enjoys learning new skills whenever possible. He lives in Atlanta, Georgia, with his wife and two young boys.

Dedication

For Decker and Sawyer, my best projects ever

Acknowledgments

I read a book on writing once that said the life of a writer is a solitary one. That statement might be true for novelists, but for technology writers it's completely unfounded. I'm fortunate to have a lot of folks to talk with and share ideas, and many of them are directly responsible for making certain my books look good and are as error-free as possible. My colleagues at Pearson continue to make writing about technology enjoyable, and I'd like to thank Rick Kughen for taking the most basic of ideas ("It's a book about 3D printing, but written for beginners who might not even know what a 3D printer is...") and letting me run with it. Along the way, I've had a great support staff of editors that include William Abner, Barbara Hacha, and Mandie Frank. Just turn back a few pages and take a look at all the names of the people involved in making this book a reality—if you like what you read, please take a moment and email them a note of thanks.

In 2012 I had the good fortune of backing a 3D printer designed by Brook Drumm and sold through his company, Printrbot. Printrbot continues to grow, and Brook has been so generous in providing me with technical assistance, hardware, software, and just plain moral support. Jeremy Gallegos is a Printrbot employee who was always available to me, and I'd like to thank him for the phone calls and email support as I built the 3D printer used in this book. Both Brook and Jeremy were amazing resources to have, and I cannot recommend Printrbot's products enough. (I'm now up to two models of Printrbot 3D printers.)

Finally, I have to thank my wife, Ashley, and my two boys. I do this with every book I write, but the sincerity behind my thanking all three of them for their support only increases with each finished writing project.

We Want to Hear from You!

As the reader of this book, *you* are our most important critic and commentator. We value your opinion and want to know what we're doing right, what we could do better, what areas you'd like to see us publish in, and any other words of wisdom you're willing to pass our way.

We welcome your comments. You can email or write to let us know what you did or didn't like about this book—as well as what we can do to make our books better.

Please note that we cannot help you with technical problems related to the topic of this book.

When you write, please be sure to include this book's title and author as well as your name and email address. We will carefully review your comments and share them with the author and editors who worked on the book.

Email: feedback@quepublishing.com

Mail: Que Publishing
ATTN: Reader Feedback
800 East 96th Street
Indianapolis, IN 46240 USA

Reader Services

Visit our website and register this book at quepublishing.com/register for convenient access to any updates, downloads, or errata that might be available for this book.

Introduction

Welcome to 3D Printing!

I'd like to welcome you to the world of 3D printing. If you're already familiar with 3D printers, how they work, and what you can do with them—well, feel free to skip ahead. I won't mind.

3D printing is exactly what it sounds like—printing something that can be picked up, held in your hands, and played with. It's 3D, meaning it's not flat like a piece of paper. It's printing because the 3D object doesn't just magically appear; it must be "printed" by a special device called a 3D printer.

All of this and much more is explained in Chapter 1, "The Big Question—What Is a 3D Printer?"—and with photos! So, if your interest is piqued and you want to learn more, feel free to skip ahead right now to Chapter 1. Again, I won't mind.

You probably want to know a bit more about 3D printing. Maybe you're a little nervous that it sounds a bit too technical, or too difficult. You'll be happy to learn that there are kids doing this 3D printing thing. Young kids. How young? My oldest boy is six, and he's learning much of what you'll learn in this book and he's having a blast. I've even heard of much younger kids designing and printing out fun little objects with a 3D printer.

What kinds of objects can 3D printer owners print? I've seen a range of objects from the simple to the advanced. Buttons, game tokens, and money clips are easy to design and print and are great examples of small, simple objects that can be made in plastic. But on the advanced side, I've seen a 2' tall Eiffel Tower, a life-size human skull, a set of working gears that were inserted into a robot to make it go faster, and even a camera shell that holds film and takes real pictures. (If you just can't wait to see what people are printing with 3D printers, point your web browser to www.thingiverse.com and spend a few minutes browsing around this library of free object files that users can download and print on their 3D printers.)

There's really no need to be intimidated by 3D printing. Yes, this is a technology book, but I promise that I've written it for a nontechnical audience.

As you get a few more chapters deeper into the book, you'll discover that I've pulled back the complicated and strange workings of this thing called 3D printing. I even picked a special 3D printer to use with this book. It's called The Simple. How cool is that? It's a small 3D printer that you can build from an inexpensive kit. But you don't have to buy it or any other 3D printer right now. Read the book to see what's involved; read my notes on building a 3D printer from a kit and testing it, and see how I created my own 3D bobbles for printing. When you're done with the book, I hope you'll find that the 3D printing hobby isn't scary or intimidating. As a matter of fact, I hope you'll be looking at 3D printers, comparing them and trying to figure out which model will work best for your needs.

So turn the page, start learning a bit more about what 3D printing is, how it is done, and what hardware and software is involved. If you decide that you want to give 3D printing a try, I can make you one more promise—you are going to have so much fun.

See you in Chapter 1!

The Big Question—What is 3D Printing?

I don't know if you've heard of the term "3D printer," but I'd be surprised if the majority of readers didn't at least have an idea about the subject. But I'm going to start from scratch here. I'm going to assume that you know nothing...zero...nada...zilch. I'm going on the assumption that you are reading this book and scratching your head and saying "Nope...no idea."

However, if you are familiar with 3D printers, you can refer to the Table of Contents and find the chapter that best fits your level of experience. Chapters 2, "Find Yourself a 3D Printer," and 3, "Assembly Assistance for the Printrbot Simple," for example, go over one very specific device (and how to build it) that I'll be using throughout this book. If you already own a 3D printer and/or have already put one together, those chapters can easily be skipped. (But they have some cool sidebars and extra information that you might not want to miss.)

I'm going to start from the beginning and explain everything to you as if we were having a simple conversation. No complex technobabble. No calculus or physics required. And certainly no power tools. Let's keep this simple, shall we? Because as you're about to learn, 3D printers are for the masses, not just for the scientists and gizmo-gadgety gurus.

What Is a 3D Printer?

Let's start with the last term first—printer. Printers come in two varieties. There's the kind that stands behind a counter in a building full of machines and takes your money when you ask for 150 copies of your son's graduation invitations or need a large GRAND OPENING banner printed for your new business. I'm not talking about that kind of printer.

I'm speaking of the other kind of printer—the one that you usually see sitting on a desk or being shared by a few dozen office workers. It's typically bigger than a shoebox and smaller than a car. Many of these kinds of printers can print in color, but not all. They can print a single sheet or the entire 350 pages of your new novel. Figure 1.1 shows a typical printer that spits out paper with ink on it.

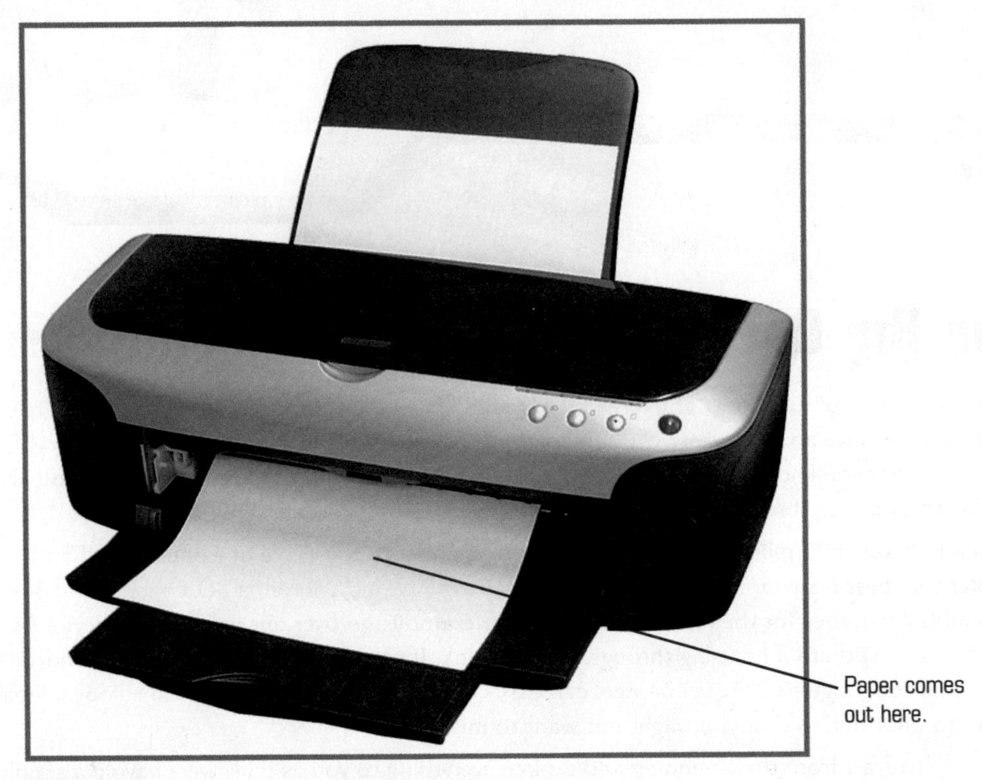

FIGURE 1.1 A conventional printer, in all its glory.

Printers print on a flat surface. Another way of looking at this is that they print in only two dimensions. Think back to your early math classes and you may remember discussions of dimensions. A one-dimensional object is nothing more than a point in space that can move along a fixed line (like the period that ends this sentence, but without any volume). A point has no height, width, or length. It's simply a point. Likewise, a two-dimensional object is flat; it has length and width but no height. Figure 1.2 shows some examples of one-dimensional and two-dimensional objects.

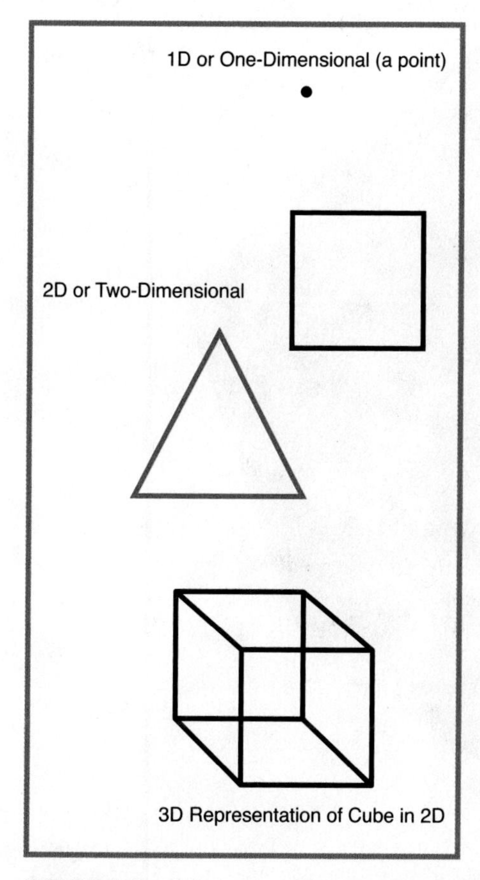

FIGURE 1.2 Examples of one- and two-dimensional objects.

Two dimensional objects can be drawn in such a way that they look like they have length, width, and height, but they're still flat. They still exist only in two dimensions.

A 3D printer is a printer that prints in three dimensions. That's as simple a definition as you'll ever find. When your inkjet or laser printer prints out a square on a piece of paper, all you have is a two-dimensional square on a flat piece of paper. It lacks height. In fact, an advertisement for a 3D printer might state, "New and improved! Now comes with HEIGHT!"

But if a 3D printer can add that *third* dimension, then it's possible to print that square as a cube, as in Figure 1.3.

FIGURE 1.3 A basic example of what a 3D printer can accomplish.

And there you have it. A 3D printer can print something with length, width, and height. But I can already hear some of you saying, "But no matter how many times I print that square on the same piece of paper, it's going to stay flat. This is crazy talk!"

And you're right.

Ink is not very useful for printing solid objects.

Say Hello to Plastic!

Think about this for a moment: You print out a black square on a sheet of paper and then cut it out with some scissors. You then place that black square on a table. Next, you print out another black square, cut it out, and stack it on top of the previous black cube. Do this 500 times. Are you seeing what's happening to that stack of black squares? Is it starting to look like a cube?

NOTE

Depending on the thickness of your paper and the size of the square, you might have to print and stack 200, 500, or even 1,000 squares to make it look like a cube. (And I don't recommend that you actually do this. Save the paper and some trees and just imagine all those stacked squares, and you get the idea.)

What you're seeing happen here is due to layering. If you could peel the ink off each piece of paper (and have it hold its shape) and then stack all those black squares like the ones in Figure 1.4, you'd (hopefully) have a solid black cube consisting of dozens, or hundreds, or maybe even thousands of layers.

FIGURE 1.4 A stack of black squares begins to resemble a black cube.

That's how a 3D printer creates three-dimensional objects: one layer at a time.

But again, I hear some of you asking, "How exactly does a 3D printer stack ink in layers?" And the answer is simple—3D printers don't print with ink. Instead, they print using a material that you are already familiar with—plastic.

Look at a two-liter bottle of soda or a kid's action figure, and you're looking at a three-dimensional object most likely made of plastic. It's a rugged material that (usually) holds its shape, is (usually) waterproof, and (usually) won't melt in the backseat of your car.

Plastic requires a very high temperature to melt, so it's a favored material for printing three-dimensional objects. And now you've learned something else about what makes a 3D printer work—a high temperature. A 3D printer must have a method for heating up plastic until it changes from solid to liquid form.

NOTE

Not all 3D printers use plastic. You'll learn about some other types of 3D printers later in the book that use different methods and materials, but for the purposes of this book, I'll be covering those kinds of printers that melt plastic to a liquid form and deposit (print) that liquid plastic on a surface so that it can cool and harden.

Solid to Liquid

Plastic is somewhat of a vague term because there are many types of plastics. One type might be much harder to twist or break, whereas another might require a higher temperature before it begins to melt. I talk about the various types of plastic used with 3D printers later in the book, but for now I just want you to understand that before a 3D printer can actually "print" a 3D object, it must melt that plastic.

To do this, most 3D printers use plastic filament. Plastic filament is nothing more than a very thin strand of plastic that typically comes in a wrapped bundle like the one shown in Figure 1.5.

FIGURE 1.5 A bundle of plastic filament, ready for melting.

Most 3D printers that print in plastic use electricity to heat up a component called the *hot end*. The hot end melts the plastic, and works with another component called an extruder that pushes the solid plastic into the hot-end. Inside the hot-end, the solid plastic melts and exits the hot-end's nozzle onto a flat surface. This is called *extrusion*. This melted plastic quickly cools and solidifies.

Figure 1.6 shows an example of both an extruder and a bit of melted plastic coming out of the bottom as a thin, fine string.

FIGURE 1.6 Melted plastic exits the extruder as a thin string of plastic.

Have you ever seen someone write Happy Birthday with icing on top of a birthday cake? The decorator squeezes the icing and moves his or her hands and, hopefully, out comes "Happy Birthday, Mom!" and not "Hopop Diaddap, Nom!"

Hold that thought. A 3D printer works in a similar manner, but without the shaky hands. The cooling plastic is applied to a flat surface, but the plastic is placed on the surface in such a way that when it cools and hardens, its shape has changed from the original loop of plastic filament (seen back in Figure 1.5).

But how does a 3D printer apply the plastic on the flat surface? Most printers don't have hands, right? Well, it turns out that 3D printers don't need hands—they have something much better.

A Different Type of Motor

Say the word "motor," and a lot of people's eyes glaze over. Motors are strange, magical things to some and intimidating and scary things to others. But motors are important. They get your car from point A to B. They keep your fridge cold (and everything inside). They make certain the light bulbs have power so you can finish reading this book. Motors are everywhere, and they come in all shapes and sizes. And yes, 3D printers have their own versions of motors.

You don't have to be an electrical engineer to understand the motors found in a typical 3D printer. I promised I wasn't going to get super-technical on you, so for now I want you to know these five things about the motors found inside a standard 3D printer:

- They are used for movement.
- They can spin clockwise or counterclockwise.
- They can spin at varying speeds.
- They require electricity.
- They are controlled by a computer.

Most 3D printers have four motors. Remember when I told you that a 3D printer could print objects that have length, width, and height? One motor each is used to move the extruder (or the printing surface; more on that in a moment). I'll explain how this movement occurs later in the book, but for now all you need to know is that the motors spin clockwise or counterclockwise to create movement. (The fourth motor is used by the extruder to feed or push the plastic filament into the hot end so it can be melted.)

What do these motors look like? Take a look at Figure 1.7 and you'll see a typical 3D printer motor. They don't all look like this one, but most consist of a hard shell and a shaft that rotates clockwise and counterclockwise.

FIGURE 1.7 A motor like this one is typically used in 3D printers.

This type of motor is called a stepper motor, and it has the unique capability to spin (clockwise or counterclockwise) in very tiny increments. This means the shaft can spin not in 1-degree increments, but in fractions of a degree! This is an important concept, and you'll get some more explanations later in the book. Right now, what you need to take from this discussion on motors is that they do the heavy work by moving items such as the hot end or the plastic filament feed.

Let's take a quick look at what we now know about 3D printers:

- 3D printers can print three dimensional objects.
- 3D printers use melted plastic to create objects.
- 3D printers use a hot-end to heat up and squeeze out melted plastic.
- 3D printers need motors to control the movement of the hot-end.
- 3D printers use four motors.

Again, 3D printers melt plastic and require motors to move the extruder around as it squeezes out the melted plastic. But how exactly does a 3D printer create objects with recognizable shapes instead of just squirting out one big blob of cooled plastic?

It's all done by carefully controlling the movement of the extruder along three paths. These paths go by different names, but you can think of them as left/right, forward/backward, and up/down. They also go by shorter names: the X axis, the Y axis, and the Z axis.

3D Objects Require Three Axes

Think back to your early math and recall that all points in space can be described by three coordinates: X, Y, and Z. If you need a refresher, take a look at Figure 1.8, which shows all three axes on an xyz graph.

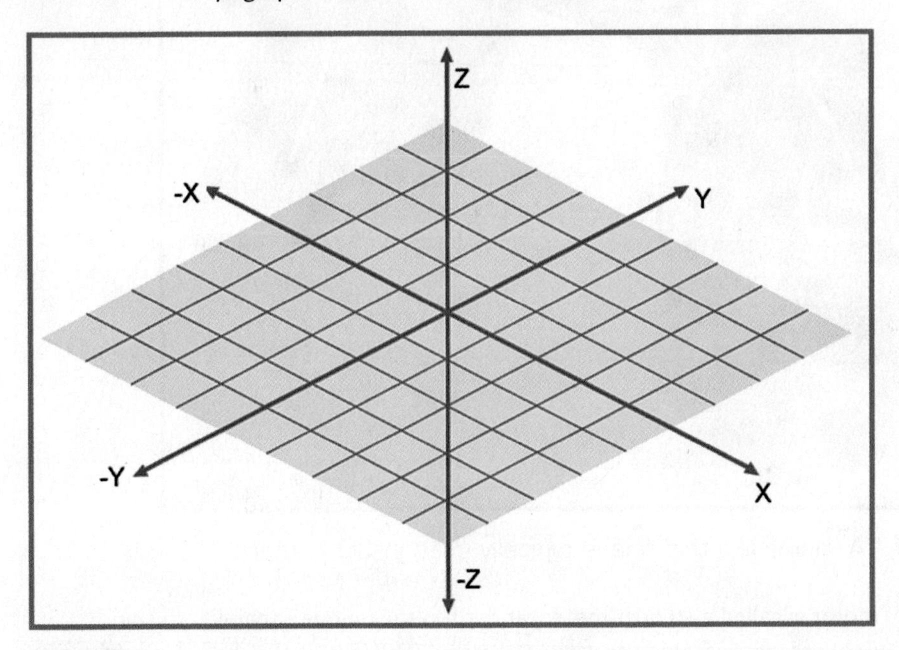

FIGURE 1.8 A point in space has an X coordinate, a Y coordinate, and a Z coordinate.

NOTE

When talking about X, Y, and Z, use the plural term "axes." The singular term is "axis." So, the extruder moves back and forth on the X axis but also can move in any direction using three axes.

You may be wondering just how the xyz graph relates to a 3D printer. It's now time to show you one example of a 3D printer. Don't worry if it looks strange or if you can't figure out how it works; that will soon become apparent. What I want you to focus on are the three directions that the extruder can move. Everything is labeled in Figure 1.9.

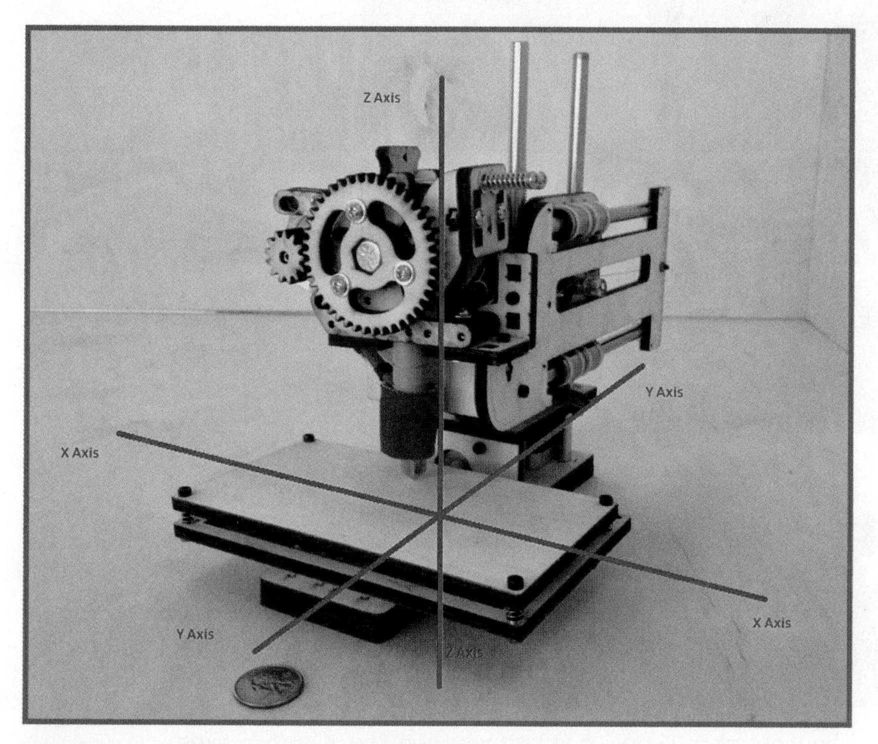

FIGURE 1.9 A 3D printer's extruder can move in three directions.

If you're trying to force the xyz graph from Figure 1.8 to fit the 3D printer shown in Figure 1.9, you might be confused. That's because Figure 1.9 is showing the 3D printer from the front. Let's take a look at the 3D printer again, but this time looking down on it from above, as shown in Figure 1.10.

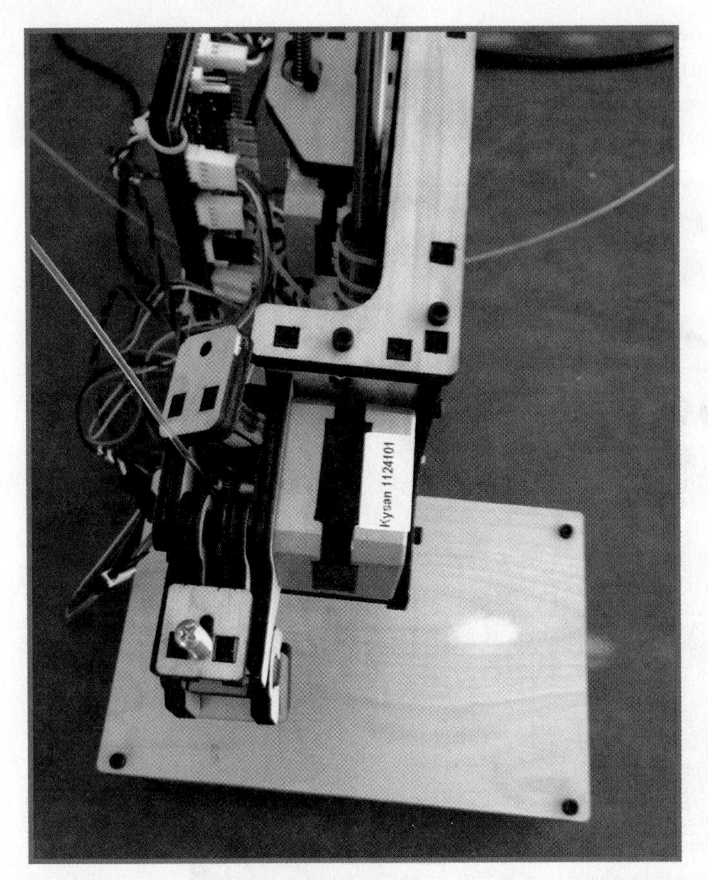

FIGURE 1.10 Top view of a 3D printer lets you see the X and Y axes.

Looking at a 3D printer from above, you can now see that when the hot-end moves left or right (when viewed from the front of the 3D printer), it's actually moving along the X axis. Likewise, when the hot-end moves forward or backward (again, when viewed from the front), the hot-end is moving along the Y axis.

Now go back to Figure 1.9, which shows the 3D printer from the front. Can you see the Z axis now? When the hot-end moves up or down, it's moving along the Z axis.

Here's another idea to mull over. If you look at the 3D printer from above, imagine that the X axis can be used to define a two-dimensional object's length (like a square). The Y axis can be used to define that two-dimensional object's width. So what do you think the Z axis will be used to define for an object? If you answered "height," you'd be correct.

A 3D printer's motors will be used to move the hot-end left, right, back, forward, up, and down. While these movements are occurring, the extruder is constantly pushing in the filament to melt in the hot-end and then coming out the hot-end's nozzle a thin thread of melted plastic. This melted plastic continues to extrude at a constant rate, almost like ink from a pen as you write your name on a piece of paper. The only difference is that as the plastic cools, it begins to harden. Any melted ink placed on top of an existing (hardened) bit of plastic will simply add to the height of the plastic at that point. And that's how we get the layering effect that a 3D printer uses to print 3D objects!

Let's return to that solid black square drawn on a piece of paper. If we use our 3D printer to create that square, what will be required? The extruder needs to move along a path that lets the melted plastic form a square that has the same thickness. Two possible paths it could follow are shown in Figure 1.11.

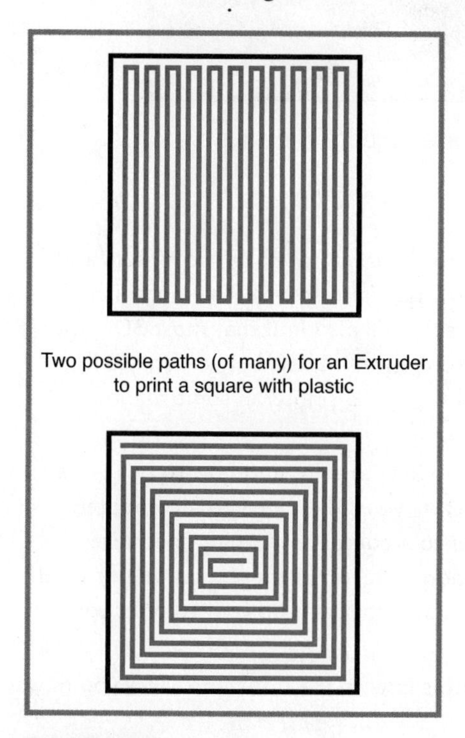

FIGURE 1.11 A single layer of melted plastic in the shape of a square might use one of these paths.

The thickness of the melted plastic is actually thinner than that of a pen (and its ink), but you should get the idea. The hot-end acts like a pen, drawing a path on the flat work surface and creating a square-shaped object. What do you think will happen if the hot-end places another layer of plastic over this first layer? What happens if the hot-end places 10 or 20 or even 50 layers, one on top of each other?

You start to see the beginning of a plastic cube.

A plastic cube might not be much to look at, but if you understand how layers of plastic are combined to create a three-dimensional object, you are well on your way to understanding how a 3D printer can be used to create more advanced objects like those shown in Figure 1.12.

FIGURE 1.12 You can use a 3D printer to create objects like these.

A Few Other Items

Most 3D printers have a small circuit board (or controller) that "talks" to the motors and gives them instructions on how to move (how far, how fast, and so on). The motors and the circuit board/controller need electricity to work, so you'll also find that most 3D printers need a power supply. This often comes directly from a wall outlet, but sometimes it requires an actual power supply box that is attached to the 3D printer in some manner. You'll see the circuit board later in Chapter 3.

A 3D printer can print 3D objects, but until you tell it what to print and how to print, it's just a paperweight. For that reason, you'll be connecting your 3D printer to a computer. More specifically, you'll be connecting the 3D printer to a computer running specialized software. This software (sometimes a single application, sometimes more than one) is used to create an object and send data to the 3D printer that instructs it on how to print your object using melted plastic.

Don't let this overwhelm you. I introduce you to all this later in the book and show you how easy it is to do. (And how much fun it can be, too!)

So, let's summarize what we know:

- 3D printers use motors to move a hot-end around.
- An extruder pushes plastic filament into the hot-end.
- Melted plastic exits the hot-end's nozzle onto a flat surface.
- Special software provides instructions to the motors that define a path for the hot-end to follow.
- The hot-end follows the path multiple times to create layers.

If you've read and understood everything so far, you're off to a good start! There's much more to learn; in the next chapter I introduce you to a very simple 3D printer and some variations of other 3D printers. You'll also read explanations on the benefits and drawbacks of different models, so you can make an informed decision should you choose to purchase a 3D printer kit.

Find Yourself a 3D Printer

If you don't yet own a 3D printer, this is the chapter for you. I give you some things to ponder when considering what type of 3D printer you might want to purchase. Believe it or not, you have hundreds of options compared to just three or four years ago. As in the previous chapter, my goal here isn't to overwhelm you with a lot of technical talk. These days, when most people choose to purchase a computer, they don't go to the store asking about memory cache amounts or hard drive transfer speeds (well, some do). Instead, they just want something that works and can run the software they need it to run. Today's 3D printers are well on their way to being that easy to purchase and use (what computer companies used to advertise as plug-n-play). You can buy preassembled 3D printers, ready to go right out of the box. Just connect it to your printer, follow some simple instructions for installing the software, and you're all set. (This is not completely true, because you've got to have something to print. More on that later in Chapter 5, "First Print with the Simple.")

There are also 3D printer kits. These require you to assemble them, but today's 3D printers have gotten much easier to figure out and put together compared to just a few years ago. Don't believe me? I build a 3D printer and document much of the assembly in Chapter 3, "Assembly Assistance for the Printrbot Simple." When you're done with Chapter 3, I hope to have you convinced that a kit should not be intimidating if you choose that route.

So, let's get started by taking a brief look at some of the options you have when purchasing a 3D printer. The goal for this chapter is to help you identify what you want to do with a 3D printer, how much you want to spend, and a few other factors. By chapter's end, you should have a better idea of which 3D printers will meet your needs.

3D Printer Options to Consider

The practice of printing three-dimensional objects has been around for years. But it's been only in the past five to six years that 3D printing has become a popular hobby and favored topic in the media, mainly due to costs dropping and component sizes shrinking. Today, 3D printing is available to anyone with a home computer and the interest to print out 3D objects.

The most popular type of 3D printing involves extruding melted plastic. You read about this in Chapter 1, "The Big Question—What Is a 3D Printer?" and that method of printing is the focus of this book. Melted plastic isn't the only method for printing 3D objects; there are a number of techniques for creating 3D objects, but only the extruded plastic method can truly be considered the "affordable" solution at this time. But it's not all about money, either.

The following list is by no means comprehensive, but it does represent the six factors that I believe someone new to 3D printing should consider when deciding on a 3D printer to purchase and use:

- Initial cost
- Ease of assembly and tech support
- Operating System compatibility
- Cost and type of filament
- Resolution/nozzle diameter
- Print bed size and leveling

NOTE

Another option that I don't cover in this book is the availability of dual extruders. Dual extruders allow for users to load up two (or more) colors of plastic filament. Printing multi-colored objects becomes available with the proper software that can instruct the Extruder on which filament to feed into the hot-end. This topic is a bit beyond the beginner level set for this book, but if you find a reasonably priced 3D printer that you like that also offers an upgrade to a dual extruder, it's worth considering as a beneficial add-on.

I'm going to go through each of these factors and explain why I believe they are key considerations for your future 3D printer purchase. For the purposes of this book, I'll limit my discussion to the plastic-extruding type of hobbyist 3D printers. (If you want to learn about other methods of 3D printing, point your web browser to http://en.wikipedia.org/wiki/3d_printing and you'll find details about half a dozen or more 3D printing techniques, such as selective laser sintering, an up and coming type of 3D printing that should become more affordable for hobbyists over the next 3–5 years.)

Initial Cost

Just a few years ago, the hobbyist 3D printer market (in contrast to the higher-end professional products used in specialty industries such as manufacturing and engineering) floated between $1,000 and $3,000. At that time, it was an amazing price for cutting edge technology that could be ordered, delivered, and assembled in your home. Still, the price was out of reach of most hobbyists, let alone any parents or teachers who might want to introduce a 3D printer to a child or a classroom.

But one amazing thing about technology (especially technology that gets a lot of media attention) is how fast the price can drop when demand goes up. Demand for 3D printers has grown, and this has resulted in hundreds of 3D printer options popping up, many almost overnight. Companies and individuals, big businesses and small businesses—the list of sellers of 3D printers is an impossible document to put together because it seems that a new 3D printer is announced almost daily. This is great news for anyone wanting to buy a 3D printer today.

The variety of 3D printers and the various options they offer have created a new market, where prices have dropped drastically. You can still find 3D printers in the high $1,000s, but these are the high-end printers that would have been out of reach of most hobbyists a few years back. And the entry-level 3D printers that floated around $800–$1,500 two or three years ago? Those can be had for under $500.

Today, you can expect to pay between $300 and $3,000 for a hobbyist level 3D printer. These are printers that are designed to be used in a home or small office, producing good to high-quality 3D objects.

But price isn't just related to the print quality of the plastic object that is created. Price is often also related to whether the 3D printer is preassembled (also referred to as ready "out-of-the-box") or comes in a kit for you to build yourself. Some 3D printers, such as the Replicator 2 from MakerBot (makerbot.com) shown in Figure 2.1, only come preassembled. The reason is that the assembly of a 3D printer often requires some fairly strict tolerances. If a screw isn't tightened properly or two panels aren't perfectly perpendicular to one another, a printed 3D object can come out a little lopsided or completely unusable.

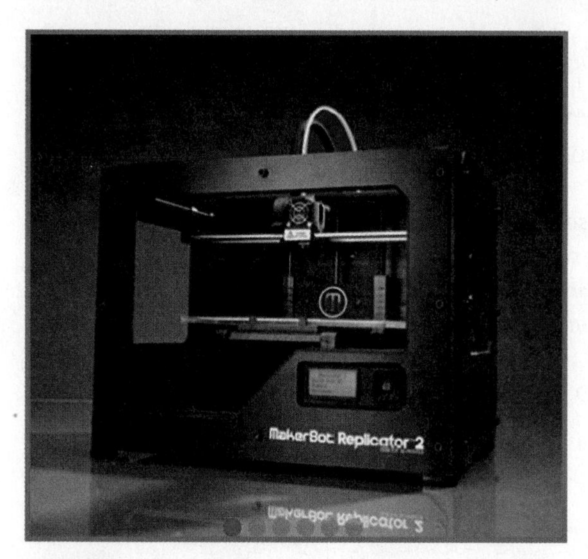

FIGURE 2.1 The Replicator 2 from MakerBot.

Other 3D printers come in both preassembled and kit form. You can typically expect to pay around $100–$300 more for a preassembled version, maybe even higher. Keep in mind that a preassembled 3D printer typically incurs a higher shipping cost. The assembled 3D printer is bigger than an unassembled version; it is most likely a bit more fragile and must be packed better to keep anything from getting damaged during delivery. Take a look at the Ultimaker website (ultimaker.com), and you'll see an example of a good mix of kit and preassembled 3D printers, such as in Figure 2.2.

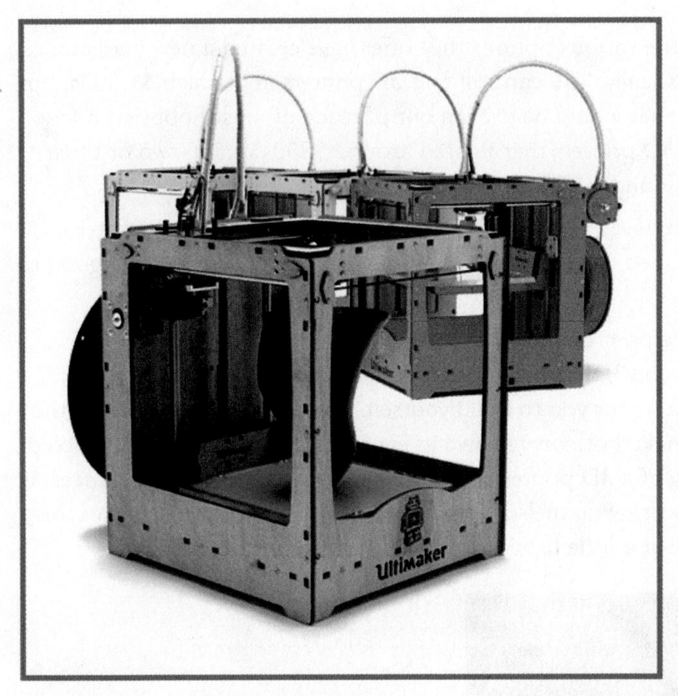

FIGURE 2.2 Ultimaker offers kits and preassembled 3D printers.

If you're feeling a bit of sticker shock at some of the prices you've seen, let me calm your nerves a bit. The 3D printer that I've chosen to use for this book won't run you $2,000. It won't even cost you $500. I use the Printrbot Simple in this book (and show you how to assemble it in Chapter 3). The Printrbot Simple comes in a kit for $300. You read that right—$300. It doesn't have all the bells and whistles that some of the mid- to high-level 3D printers possess, but it does print out 3D objects in plastic and is a great way to introduce yourself to the hobby.

NOTE

Ready to buy? Point a web browser to printrbot.com and search for the Printrbot Simple. The website may change its appearance from time to time, but right now the Simple is advertised on the front page. As Printrbot introduces new products, you may find that you'll need to look at the company's complete listing of products to find it. And who knows? The price may have even dropped by the time you read this.

The Printrbot Simple is easy to assemble, as you'll see in Chapter 3. I chose the Simple as the 3D printer to use for this book because of its low price and easy-to-understand assembly instructions that make it a great 3D printer for beginners. You can see the completed Printrbot Simple in Figure 2.3.

FIGURE 2.3 The Printrbot Simple.

Be aware that the Printrbot Simple does have some limitations. For example, the Simple can't print large objects. Imagine a cube that is four inches (4") square. That's the print limit for the Simple. If you want to print anything that won't fit inside a cube with 4" sides, you'll need to consider a 3D printer with a larger print area. But if you're wanting to experiment with the 3D printing hobby without spending a lot of money, the Simple will get you some hands-on time with both the hardware and software and let you test the waters before you consider upgrading to a more expensive and more feature-filled 3D printer.

NOTE

There's a great way to print larger objects with a 3D printer like the Simple that doesn't have the largest print area, and it involves breaking up your 3D models into smaller parts that can be connected with adhesives or methods, such as slots or even small screws. This can add to the complexity of designing an object, but you should be aware that a small print area doesn't necessarily limit you to printing larger plastic objects...you've just got to be more creative in your solution.

What might you want to look for in more expensive 3D printers? Here are some things to consider before making a purchase:

- Multicolor support—Most hobbyist 3D printers can print using one feed of plastic filament. Because filament can come in colors, there are also printers that can be loaded with two to four colored filaments using dual extruders, allowing you to print 3D objects using a mix of colors. It's a bit more advanced to print with multiple colors, but it is possible with some higher-end 3D printers. (Rumor has it that Printrbot is working on an upgrade to the Simple that will allow you to buy an upgrade for the single extruder that will turn it into a dual extruder, allowing you to print in two colors! It's not available right now, but I suspect it could be available by the time you're reading this book.)
- Customer support—Most 3D printer sellers are more than happy to answer technical questions from customers, but many of them push customers to an online forum or an email support system. Some sellers have customer support phone numbers and others do not. You'll want to consider the level of support provided when purchasing a kit or a preassembled 3D printer. Kits seem to typically spur more questions from owners, so make certain if you buy a kit that you can get help from the seller should you have problems.
- Proprietary filament—Some 3D printer sellers require you to purchase their plastic filament. They can do this because they use some type of special spool or loader that fits inside the 3D printer. This can be a blessing or a curse, depending on your level of technical expertise. Filament can often be purchased on sale or at extremely low prices from specialty companies that cater to 3D printing hobbyists, so being locked into using a proprietary filament could be more expensive over time or at least prevent you from taking advantage of price breaks on filament.
- Hot-end options—Some hot ends/extruders require you to assemble them, involving some hands-on time with a soldering iron. This can be extremely intimidating, especially if you don't even know what a soldering iron is or how it is used. Make certain that when you purchase a kit, the hot end and extruder are ready to go. You'll also want to inquire about whether you can upgrade your hot end. A hot-end can have different nozzle diameters, allowing for different qualities of printing.
- Warranty—Most kits aren't going to come with a warranty. But preassembled kits should have some sort of return policy if you find something is damaged or not working properly. Be certain to inquire about the return policy on any 3D printer you purchase, especially when buying a kit.

As you learn to use a 3D printer and the software tools needed to create 3D objects, you'll begin to understand some of the benefits offered by higher-cost 3D printers, especially those with more advanced electronics. But if you're just getting started, you'll find that any sub-$700 3D printer will offer you a great experience.

Ease of Assembly and Tech Support

If you end up purchasing a preassembled kit, you'll skip this step. Yes, you may have some instructions on connecting it to your computer or inserting the plastic filament, but for all practical purposes, the preassembled 3D printer shouldn't require much from you in terms of putting it together.

But a kit? A kit can be a bit overwhelming the first time you open it up and examine all the parts that come inside. Take a look at Figure 2.4 to see the parts used to build the Printrbot Simple.

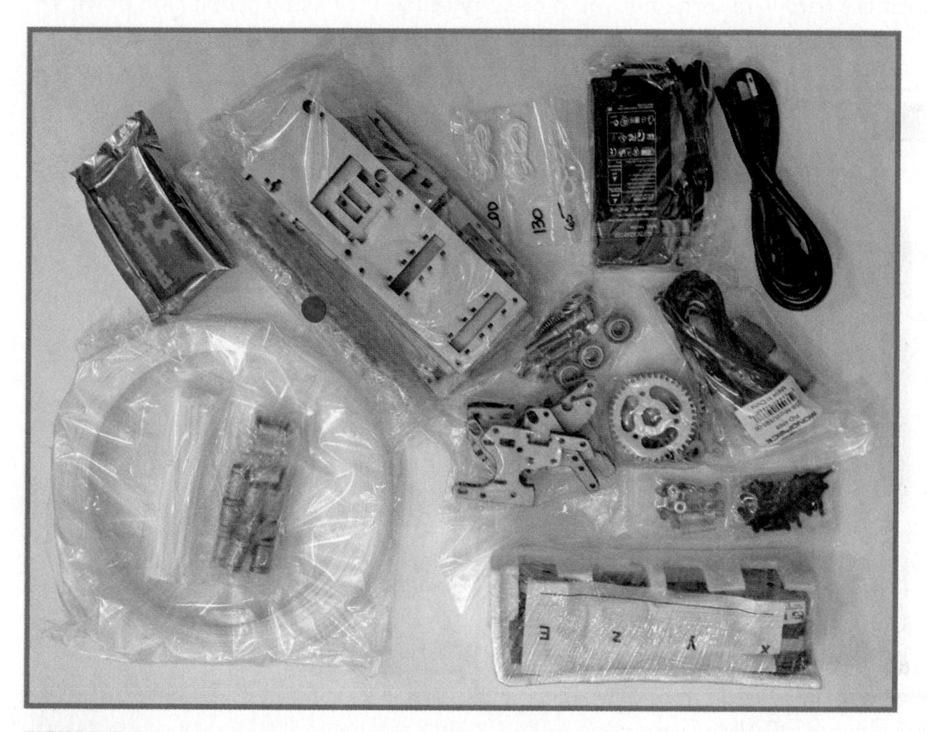

FIGURE 2.4 The unassembled Printrbot Simple.

Keep in mind that the Printrbot Simple is exactly that—simple. It has a limited number of parts required to build it (see Chapter 3). Now imagine a larger, more advanced 3D printer; it's likely to have even more parts.

Some 3D printers come with instructions tucked into the box for assembly, whereas others have the build instructions available online. My personal preference is online building instructions. They typically come with full-color photographs; errors are typically corrected as soon as they are found, and corrections are posted immediately. However, you may have only a printed manual with your 3D kit, so be aware that assembly instructions vary from kit to kit. It never hurts to ask about what types of assembly instructions are available when ordering a kit!

Tech support is also a big deal. Find out if your 3D printer seller offers a forum where you can post questions. Most do, but not all. It's great to be able to log in to a company's online forum and post a quick question along with a photo or two if necessary. You'll often find that fellow kit owners are fast to respond and offer help, often faster than the in-house technical support folks.

If you're comfortable using a screwdriver or wrench, you should be okay putting your 3D printer together. Most 3D printer assemblies require few other tools, but it never hurts to ask. For example, be sure to ask if you need an Allen wrench. This is a cheap and easy tool to pick up at the hardware store, but you'd be surprised at how many people don't own a set. See the Allen wrench set in Figure 2.5 if you're not familiar with them.

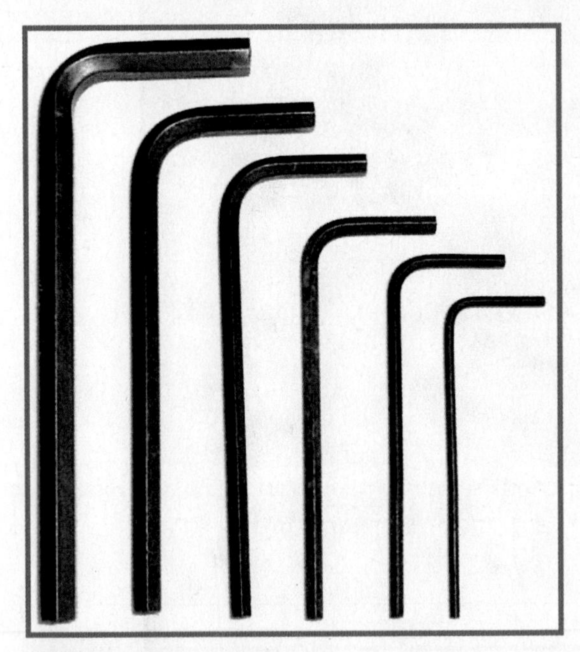

FIGURE 2.5 A set of Allen wrenches.

Consider your own comfort level, too. Not everyone enjoys taking a bag of 50 or even 100+ parts and putting it all together. I've put together three 3D printers over the past few years, and I'm comfortable reading instructions, asking questions, and "figuring it all out." But if that's not your idea of fun, a kit may not be for you. (I do believe that putting something together, like a 3D printer, is one of the best ways to understand how something works and how to repair it later.)

Operating System Compatibility

Some 3D printers work with the Apple OS X operating system (OS), and some 3D printers play well only with Microsoft Windows. Let's not forget that the Linux operating system is enjoyed by many people around the globe, but unfortunately not every 3D printer can be connected to a computer running this particular OS.

Whatever your choice of operating system, make certain that your new 3D printer can be connected to your computer. Without a compatible computer (and OS), you won't be able to print your 3D objects. The question to always ask (or look for on product pages) is "what operating systems support this 3D printer?"

By all means, dig deep and make certain that the version of your OS is supported. If you've got an older computer running Windows XP and that new 3D printer you purchased supports only Windows Vista or newer, you're going to have to upgrade your OS (if you even can) or find a new computer to connect it to.

NOTE

Be aware that it's not just the 3D printer's compatibility with your OS that you should be concerned with. You also want to make certain that whatever 3D design software you want to use will run on your OS. Software is discussed later in the book, but for now know that some 3D printing software runs only on Windows, for example. Other 3D printing software might work with only the Mac OS. (And poor Linux has even fewer options.)

Cost and Type of Filament

Because this book is focused on plastic 3D printing, you'll typically be purchasing your plastic filament in one of a half-dozen varieties. There are two types of filament, however, that you'll find most used in the hobbyist 3D printing world—PLA and ABS.

PLA is Polyactic Acid and ABS is Acrylonitrile Butadiene Styrene. I'm not a chemist, but I can tell you that these two types of plastic do have slightly different characteristics, and you need to ask which type of plastic is recommended for use with your 3D printer.

PLA has a lower melting temperature than ABS (but the hot end will still be too hot to touch), it's biodegradable and the smell is not nearly as offensive as ABS. ABS, however, produces a much harder 3D object and can sometimes be found on sale at a lower cost than PLA. PLA and ABS come in a variety of colors, as well.

Both are typically sold by the kilogram. My first 3D printer used PLA, so I'm a bit fond of it. It smells better when melted, and I like the fact that it's better for the environment.

Pricing varies from site to site, so it pays to shop around. To give you an idea of pricing, printrbot.com is currently selling one kilogram of natural color PLA or ABS at $30, but MakerBot is selling it for $43.

Resolution/Nozzle Diameter

Keep in mind that 3D printing technology changes constantly, so you'll also want to be looking into what level of resolution a 3D printer can provide. Currently there are two popular filament sizes (in terms of diameter) you can purchase. These two diameters are 1.75mm and 3mm.

I mentioned in Chapter 1 that the hot end is responsible for heating, melting, and extruding the molten plastic onto the print bed (surface). You always need to check to see what size of filament is used by your 3D printer so you can purchase the right diameter.

But the nozzle diameter is also its own consideration. The Simple extruders use a 0.4mm diameter bead of plastic and provides smoother surfaces (generally speaking) than a nozzle with a larger diameter. Think about it in terms of a fine-point pen versus the wider point of a highlighter. With the finer point, the output on paper is a finer line and sharper details. With the wider point, the output is a wider line and less detail on things like curves. When your 3D printer lays down lines of plastic on the print bed and builds up layer after layer, that 0.4mm nozzle gives smoother transitions between layers. Figure 2.6 shows the difference in surface smoothness between two nozzles of different resolution.

FIGURE 2.6 Note the finer resolution of the layers.

More and more 3D printers are coming that use the 1.75mm filament and the 0.4mm (or even smaller) diameter nozzle, and the benefits are definitely worthwhile: faster feed rate (the filament can be fed into the extruder and then into the hot end at a faster speed), more detailed designs can be printed, and the motor has to apply less power (torque) to feed the filament into the hot end. Still, the 3mm filament shouldn't be discounted. A lot of prototypers and hobbyists like a larger nozzle because it puts down more plastic in a single pass, and it takes less time to print large objects (fewer layers). Keep all this in mind as you do your shopping, but remember that filament in either diameter is just about the same price. Buy a 3D printer based on what your printing needs are. If you're designing small trinkets that might be used to create molds for jewelry, you'll probably appreciate a smaller diameter nozzle and the detail it can provide. But if you're printing out large gears for testing in your handmade robots, for example, a larger diameter nozzle will get you those gears faster.

Print Bed Size and Leveling

All 3D printers lay down the bead of melted plastic on the print bed. This print bed is a flat surface that helps to ensure that each layer is parallel to the floor and that the hot end isn't printing out a lopsided object. Most 3D printers come with a method for leveling the print bed, but not all. It's definitely helpful to be able to level the bed (and the printer itself), especially if you get in the practice of toting your 3D printer with you to various places. You never know where you'll be setting the printer, so being able to make certain the printer and the print bed are level to the ground is a nice benefit.

NOTE

If your 3D printer doesn't come with small feet for leveling, you can easily add them. Just purchase a set from any hardware store, but be sure to buy the kind that you can turn with your fingers to raise and lower. Investing in a tiny level can also be helpful when fine-tuning the printer.

Another consideration when it comes to the print bed is the bed size print area. The Printrbot Simple, for example, is a smaller 3D printer, and it can print out only objects that would fit inside a 4" square cube—or 64 cubic inches. This might not be a problem for someone wanting to print out custom game pieces for Monopoly, for example, but if you're thinking you want to print out a 6" tall robot body, you'll want to find a 3D printer that can handle the printing of larger objects or resort to breaking that robot up into individual parts (arm, body, head) and assembling them after printing. For example, the MakerBot Replicator 2 can print objects as large as 410 cubic inches—that's about 11" long × 6" wide × 6" tall.

You'll find printers that can print extremely tall objects but are limited to length and width. Likewise, you'll find printers that can print extremely wide and long objects but don't offer much in terms of Z-axis movement (height). Again, think about exactly what kinds of objects you want to print. If you're not sure, look for a happy medium or just go with the inexpensive Simple to see what you can do. By the time you find that the 4" cube printing area isn't big enough, you'll have a better idea of what you want to create with your next 3D printer purchase. (You'll quickly find, as many 3D printer owners have, that one 3D printer is never enough. It's an addictive hobby; I own three 3D printers myself.)

Do Your Homework

I would place researching and purchasing a 3D printer somewhere between the purchase of a toaster and a car. You've definitely got a number of factors to consider, but don't let them overwhelm you. Your first 3D printer doesn't have to be the one with the most bells and whistles, and truth be told, unless you're extremely comfortable with technology and troubleshooting, you'll likely find yourself getting frustrated and overwhelmed with the most advanced 3D printers out there.

Instead, start as small and as uncomplicated as you can get. There's a very good reason I've chosen the Printrbot Simple for this book—actually, there's a number of good reasons:

- Its low cost—At $300, it's a perfect way to discover whether the 3D printing hobby is for you.
- It's easy to assemble—Compared to my first 3D printer that took me almost an entire weekend to assemble, the Simple took somewhere between two and three hours. (I took plenty of breaks and worked slowly to avoid mistakes.)
- It doesn't use a proprietary filament cartridge.
- It comes standard with the finer 0.4mm hot end that also uses the 1.75mm filament.
- The Printrbot crew has proven to me that they have outstanding customer service. (This is my second 3D printer from printrbot.com.)

The Simple comes with an excellent set of assembly instructions, and in the next chapter I provide commentary and photos on building it.

But should you purchase another brand of 3D printer, you'll still find the rest of the book useful. The 3D printer is the hardware, but there is still a software element that you'll need to understand, and my goal with this book is to make this technology as easy to understand as possible.

So, let's continue to Chapter 3, where I use the assembly of the Printrbot Simple to explain more concepts related to printing 3D objects.

Assembly Assistance for the Printrbot Simple

If you don't own a 3D printer but are in the market for one, you may be wondering about the pros and cons of purchasing a kit versus a preassembled printer. I've built two 3D printers, and I can tell you that it was fun putting all the parts together, tightening the bolts, attaching the motors, and wiring up the electronics. But I'm comfortable doing those kinds of things. You may not be.

There is nothing wrong with purchasing a 3D printer that comes ready to use, right out of the box. I'm not comfortable doing my own taxes or dealing with plumbing issues. I know my comfort level with many things, and taxes and plumbing are on the HIRE SOMEONE list.

Most 3D printers come with excellent documentation. Questions you may have can usually be answered with a phone call or email to the company, but many 3D printer hobbyists out there are also quite friendly and helpful to novices who post questions on a forum. Many 3D printer kits may look complicated as you stare at all those bags of loose components, but I know a number of nontechnical folks who have successfully put together a 3D printer by moving slowly and steadily through the process.

To give you a better idea of what's involved with a 3D printer assembly, I provide some commentary that goes hand in hand with the assembly instructions for the Printrbot Simple. The Printrbot company has great instructions on its website, and I'm not going to re-create the entire assembly document it has already created. Instead, I include some photos of my Simple's assembly and use the various stages of completion to provide you with some more details and discussion about 3D printers and areas of assembly that might give you pause or concern.

Before you start this chapter, however, you might want to open a web browser and point it to the following address:

http://printrbot.com/shop/printrbot-simple/

This is the Simple's home page, where you can find the online and up-to-date building instructions provided by Printrbot. Because Printrbot occasionally makes improvements to its kits, you should check the official website for the most accurate assembly instructions. If you're feeling ambitious, feel free to read along with my assembly discussion while you view the official Printrbot assembly instructions online.

While you're there, be sure to download the Bill of Materials (BOM) PDF file so you can get a complete listing of all hardware included in the kit. Should you purchase the kit (versus the preassembled version), you can use the BOM to verify you've got all the parts you need to do the assembly.

Printrbot Simple Assembly Part I

The first thing I'm going to do is verify that I have all the parts. I mentioned in the previous section that the Printrbot website offers a PDF file for download that contains the BOM. I can use this to make certain I have all the nuts and bolts, laser cut pieces, motors, and so on. You get the idea.

Figure 3.1 shows the current version of the BOM. Don't go by this one because Printrbot may occasionally update the Simple and change the parts list. But as you can see, it lists things such as qty-12 8mm Linear Bearings, qty-34 Hex Nut- M3, and qty-1 Extruder gear, among others. (Keep in mind, however, that the BOM for your Simple may very well have changed by the time you're reading this.)

☐		Bearings/Zip ties	☐		Hardware
	12	8mm Linear Bearing		1	2 1/2" 6-32 screw
	43	Zip Tie		1	Hex Nut - 6-32 (locking)
☐	1	LaserCut Simple Parts Kit		18	M3 10mm screw
				30	M3 16mm screw
				6	M3 20mm screw
☐		Rods		4	M3 30mm screw
	2	10" Smooth Rod		34	Hex Nut - M3
	4	6 1/2" Smooth Rod		1	M5 16mm screw
	1	7 1/8" Acme Rod			
				1	Acme Nut - 1/4"
☐		Wiring Bag			
	1	Hotend Power Cable		4	Bed Springs
	1	Hotend Thermistor Cable		1	Spring - 3/4"
	1	red/black power adapter		1	Extruder Gear
	1	Micro USB Cable		1	625ZZ Bearing
	1	Ubis Hot End (1.75mm, 0.4mm tip)			
	1	3ft fishing line	☐		Motors
				4	small Kysan motors
☐	1	Laptop-style power supply			
☐	1	Filament	☐	1	Electronics Board

FIGURE 3.1 The BOM for my Printrbot Simple.

After I've verified that I have all the parts, it's time to start assembly. Some 3D printer manufacturers provide the assembly instructions in printed form, and others may send a DVD that contains documentation or videos that walk you through the assembly. Printrbot puts its assembly instructions online; you can see the first few steps in Figure 3.2.

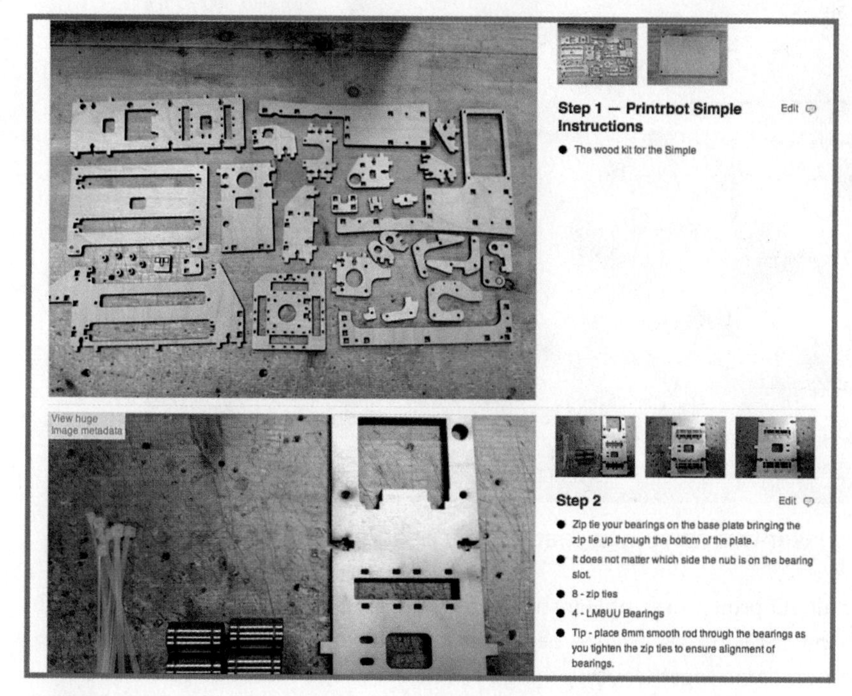

FIGURE 3.2 The first assembly steps for the Printrbot Simple.

To make things go a bit smoother, I've laid out all the laser-cut wooden pieces to make them easier to find. As you can see in Figure 3.3, my layout of parts matches the layout for step 1, shown back in Figure 3.2. The only difference is that I've added the rectangular print bed near the top of my layout.

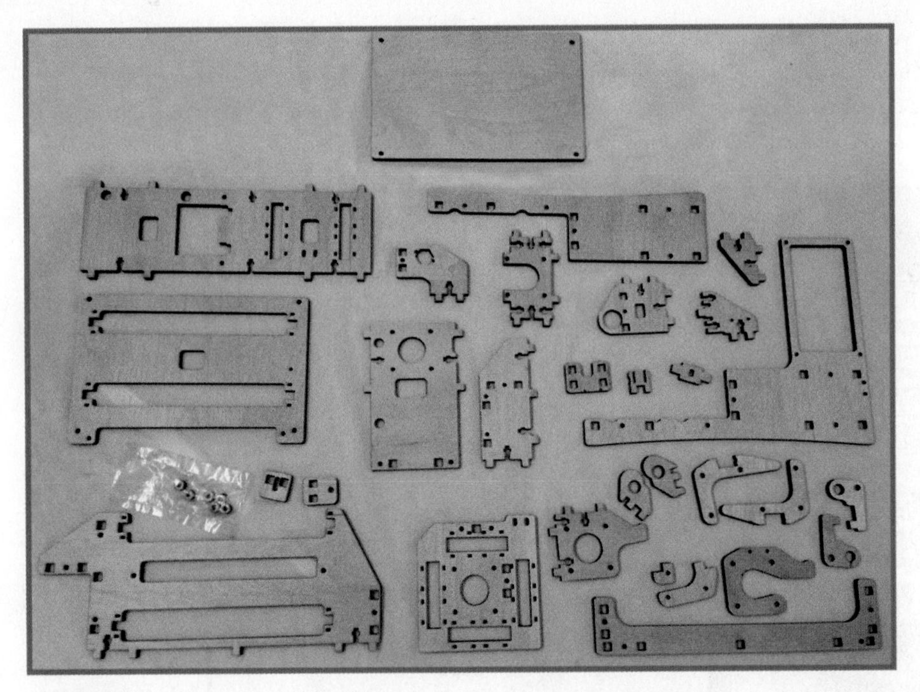

FIGURE 3.3 My laser-cut parts are ready to be assembled.

I can't speak for all 3D printer assembly instructions, but I have found that reading through the instructions one or two times prior to beginning assembly often helps to visualize how everything is going to come together. One thing I like about the Printrbot Simple instructions is that there are multiple photos for each of the steps. If you move your mouse pointer over another photo, the main image changes. Doing this can often help you to verify that you've got a part oriented properly when connecting to other parts.

NOTE

You might want to lay out all your wooden pieces and try to match them up to the ones shown in the official assembly instructions. It's a great way to make sure you have all the wooden parts before you begin assembly.

Early Assembly Observations

In the first few steps, you assemble the base of the Simple. You see how to connect bearings to one of the larger laser-cut pieces. The bearings have a slightly greasy feel because some oil has been applied to them for lubrication. Later you'll insert metal rods through the centers of the bearings, and the lubrication will allow them to slide back and forth easily. Figure

3.4 shows that I've got all four bearings attached to the wooden piece with zip ties. Notice that I haven't yet clipped the zip ties like the ones shown in step 2 of Figure 3.2. I'll clip them before moving on to the next step.

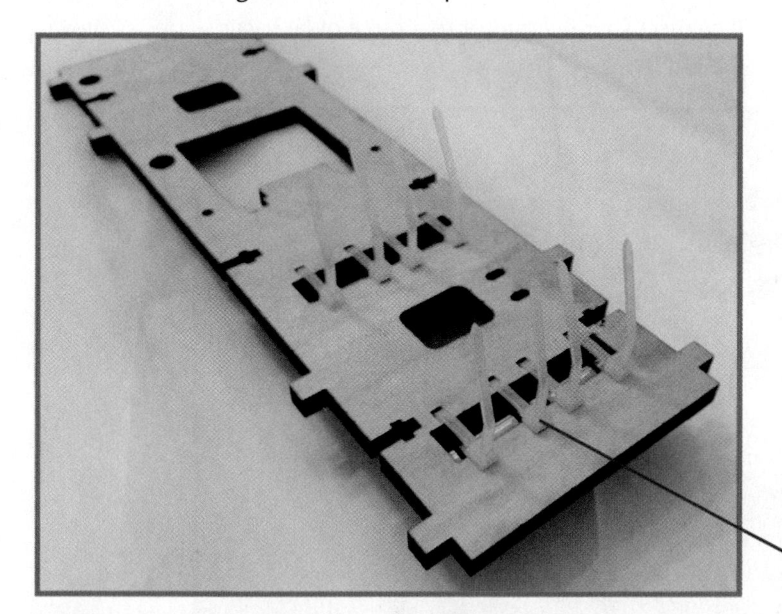

FIGURE 3.4 Metal rods will be inserted into bearings (underneath the wood piece).

NOTE

If you look back to Figure 3.2, notice that the BOM describes the bearings with a part number—LM8UU. Fortunately, all the bearings included in the Simple kit are labeled. You have to look carefully, but the part number is visible on the rubber gasket that's inserted inside one end of each bearing.

Many 3D printers use aluminum rails or other materials to form a base. Instead of zip ties, these types of materials often require nuts and bolts to lock everything down. Part of the low cost of the Simple is found in the materials, such as the wooden frame pieces and the zip ties. Be certain to do your research so you'll know the method used for assembling your own 3D printer.

Speaking of zip ties, you want to get into the habit of carefully examining the photos of the assembly instructions. You'll often find clues about the proper placement of the nub that is formed when the zip tie is closed and locked. This nub can sometimes get in the way of moving parts. For example, if the text doesn't tell you which side of a piece of wood that the zip tie's nub should be on, consult a photo. During my assembly, there was never any question about which side of a piece of laser-cut wood the nub would go, but work slowly and always verify with a photo if you can.

Early in the assembly, you'll connect your first motor to a piece of laser-cut wood, as shown in Figure 3.5.

FIGURE 3.5 X-axis motor attached to a laser-cut piece of wood.

You should be aware of a few items when bolting a motor onto a piece of wood:

- You want to find and use the proper length bolt that will go all the way through the wood and into the motor. The motors for the Simple have four holes for bolts.
- You do not need to crank the bolts down tightly; tighten them with an Allen wrench until they're snug, but don't tighten so much that the bolt cuts into the wood or crushes it.
- Use a ruler (see Figure 3.6) to verify the length of all the bolts and pieces you'll be using. Many of the small M3 bolts that come with the Simple differ in length only by 5mm. Even better, open the bag of bolts and separate them into cups or baggies by size. It'll save you frustration when you're in the middle of your build.
- Each motor has a length of wires coming out of one side. Use photos to help you determine the proper orientation of the motor so that the wires are pointing in the right direction. This will help later when you're putting the Simple together and all the wires are running in the same direction (toward the circuit board). Also, many of the laser-cut pieces of the Simple have holes for routing wire through them. Use the instructions and the photos to discover the correct mounting position. (If you make a mistake, unbolting a motor and rotating it takes only a minute.)

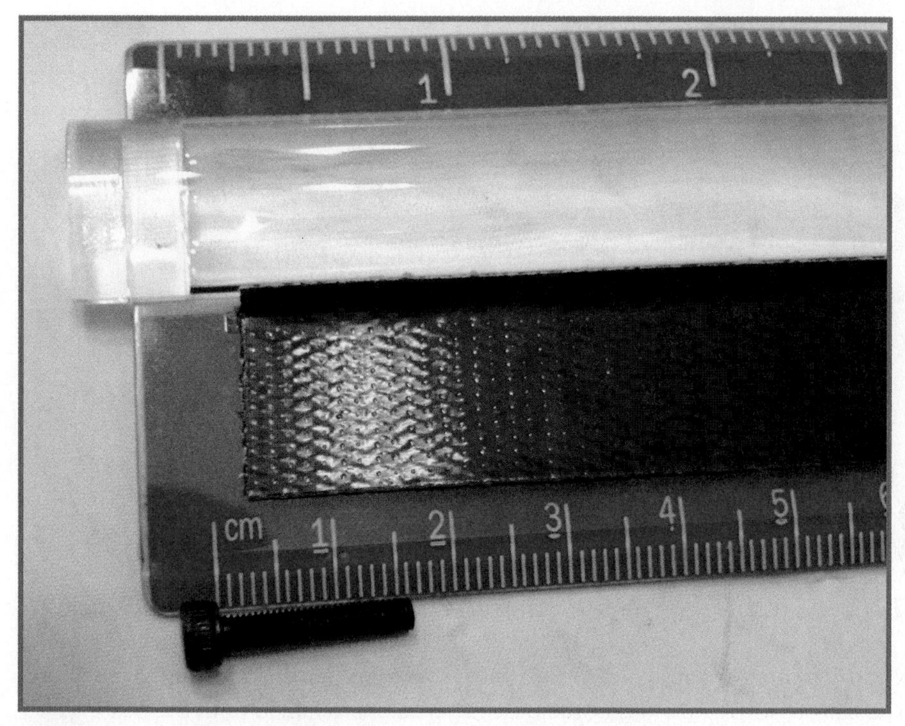

FIGURE 3.6 Use a ruler to verify the lengths of a bolt.

The first two motors you'll attach are for the X axis and the Z axis. The X-axis motor is used to move the print bed left and right (when looking at the Simple from the front). The Z-axis motor turns an item called a lead screw. Notice in Figure 3.7 that the Z-axis motor has its wires going through the precut rectangular hole. Also note that the instructions point out the two small circles that run up the left edge of the laser-cut piece of wood. This bit of information will help you attach the Z motor properly. If you mount it on the wrong side, the Z-motor axle (the metal rod that you can turn with your fingers) will have the two small circles on the right side.

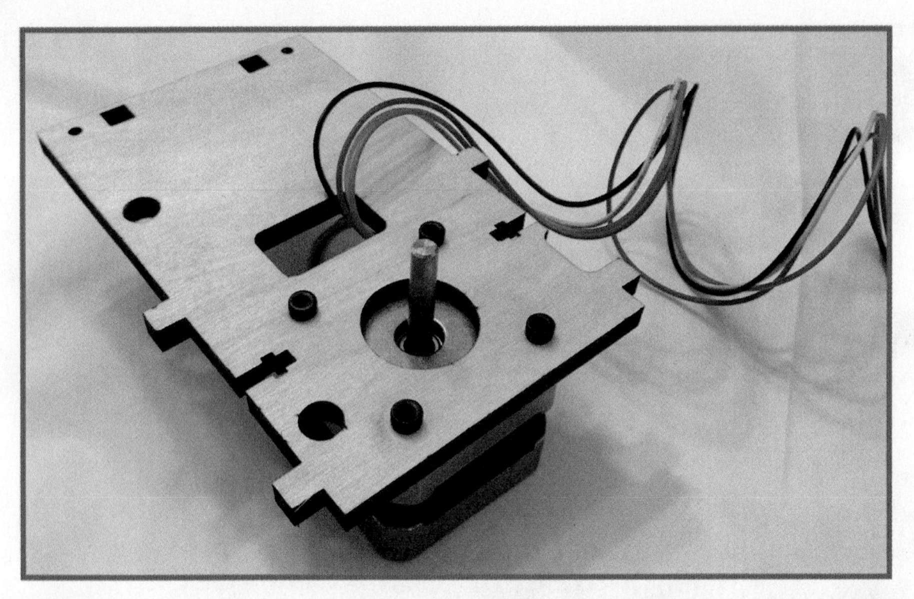

FIGURE 3.7 The Z-motor's wires are inserted up through the rectangular hole.

After the X and Z motors are attached to their respective wood pieces, it's time to assemble the base. You can see in Figure 3.8 that the base is partially assembled (and the bearings that were slightly hidden in Figure 3.4 are now visible); it still needs the right side (if viewed from the front) and the two 10" metal rods inserted into the two laser-cut circles.

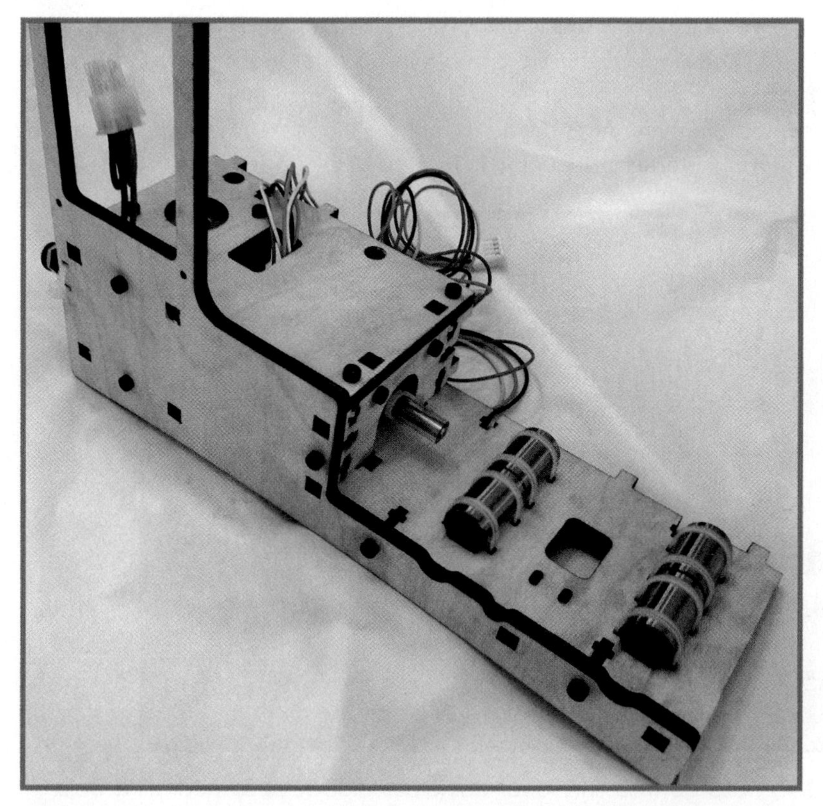

FIGURE 3.8 The Simple's base is ready for more parts to be added.

The X-axis motor and the Y-axis motor both have a small piece of clear vinyl tubing that is inserted over their axles. Don't push the vinyl tubing all the way into the axle hole, because the tube will rub and will provide resistance to the motor when it wants to spin. After you've inserted the tubing over the axle, put a zip tie on it near but not touching the motor. I put mine on with about 1/8" space of tubing left behind the zip tie, and both motors rotate just fine. Pull that zip tie as tight as you can get it, too. You'll hear a click as it tightens up, and you can use a pair of pliers (pulling on the loose end) and a screwdriver (placed against the nub) to get one or two more clicks of tightness.

NOTE

Some Simple owners have suggested putting on the zip tie first and then tightening it down before inserting the metal axle. This might work, but there's also the chance that you'll tighten too much and never get the vinyl tube to fit over the axle. When the tube is on and tightened, rotate it with your fingers. If the metal axle rotates, you've probably got the vinyl tubing tight enough.

When you connect the X-axis assembly to the base, you'll also get your first experience in using the small M3 nuts. These nuts are inserted into the small slot in the laser-cut wood, as shown in Figure 3.9. The nuts fit into the slot only one way, but even then you may have to give it a good squeeze to fit in there. After the nut is inserted into the slot, you'll screw in an M3 bolt to hold the motor (or other pieces of wood) in place.

FIGURE 3.9 The M3 nuts go into the laser-cut slots sideways to accept a bolt.

Finish the base by attaching the right-side wood piece, and make certain the Z-motor and X-motor wires are coming up out of the rectangle cut in the center of the top piece of wood.

Midway Through Assembly Observations

When it comes to the two 10" metal rods, you need to be very careful when inserting them through the cut holes. The instructions recommend using a hammer or a rubber mallet, and I highly recommend the rubber mallet option. The keys to this step are to hold each rod perfectly vertical as you hammer it into the circular hole. You're going to feel resistance because the holes are slightly smaller in diameter because the rods must be held securely and tightly for the Z-axis to work properly. You can see in Figure 3.10 that I was successful in getting both metal rods inserted; pay attention to the instructions that tell you to place some extra wood pieces underneath the bottom of the base so it doesn't crack.

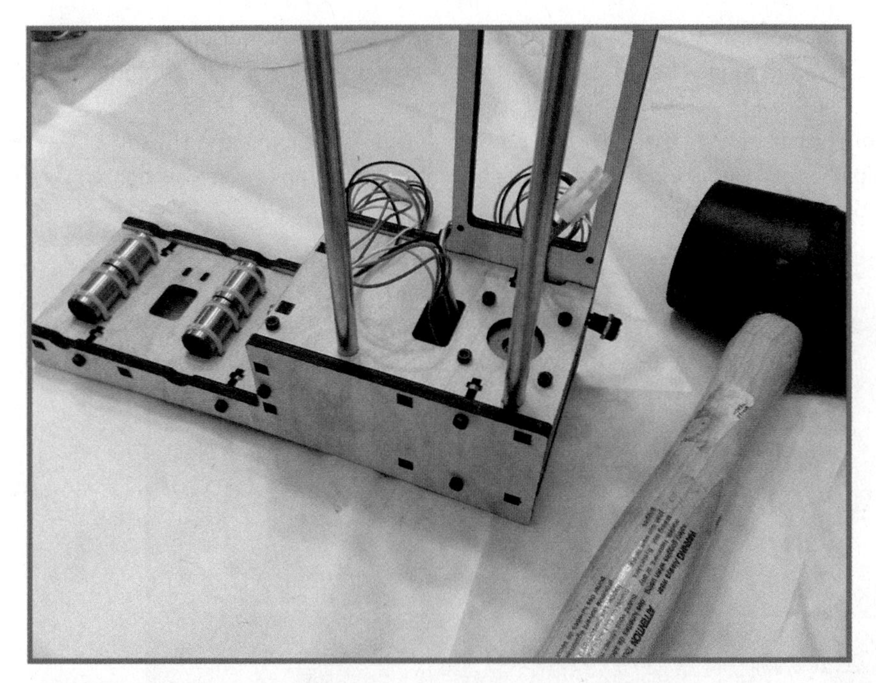

FIGURE 3.10 The Z-axis metal rods are inserted.

Next, use zip ties to connect the bottom X-axis plate shown in Figure 3.11 to the two 6.5" metal rods inserted into the X-axis bearings. These rods should move left and right easily, with almost no resistance. Pay attention to how the nubs of the zip ties rest on top of this plate and not the bottom. You also want to make certain the small hole on the plate is on the left (when looking from the front). This hole will be important if you decide to install the limit switch packet that comes with the Simple.

NOTE

What are limit switches? Limit switches will cut power to a motor when they are triggered, and they're used by the Printrbot Simple to stop the motors spinning when certain limits are reached. For example, a limit switch attached to the print bed can be used to stop the X-motor when the print bed moves as far as possible to the right. Later you can use limit switches to "talk" to the software (see Chapter 5, "**First Print with the Simple**")—this lets the software know when the motors have reached their maximum safe travel distance, as well as assist in defining a starting point (called Home) for print jobs. Note that limit switches were not available to me at the time I assembled my Printrbot, but instructions for installing limit switches are now available from Printrbot. Limit switches are also shipped with all Printrbot kits, so there's really no reason to not install them.

After you attach the bottom X-axis plate, it's time to add the fishing line used to move the X-axis print bed left and right. The trick for the X-axis is to get the line good and tight with three to four wraps around the vinyl tubing. It took me a couple of tries, but I was able to get it tight enough that moving the plate by hand would spin the tubing and the motor axle. Get a helper if you need someone to pull tight on the fishing line while you tighten the bolts. I wish I'd had someone nearby when I was at this step!

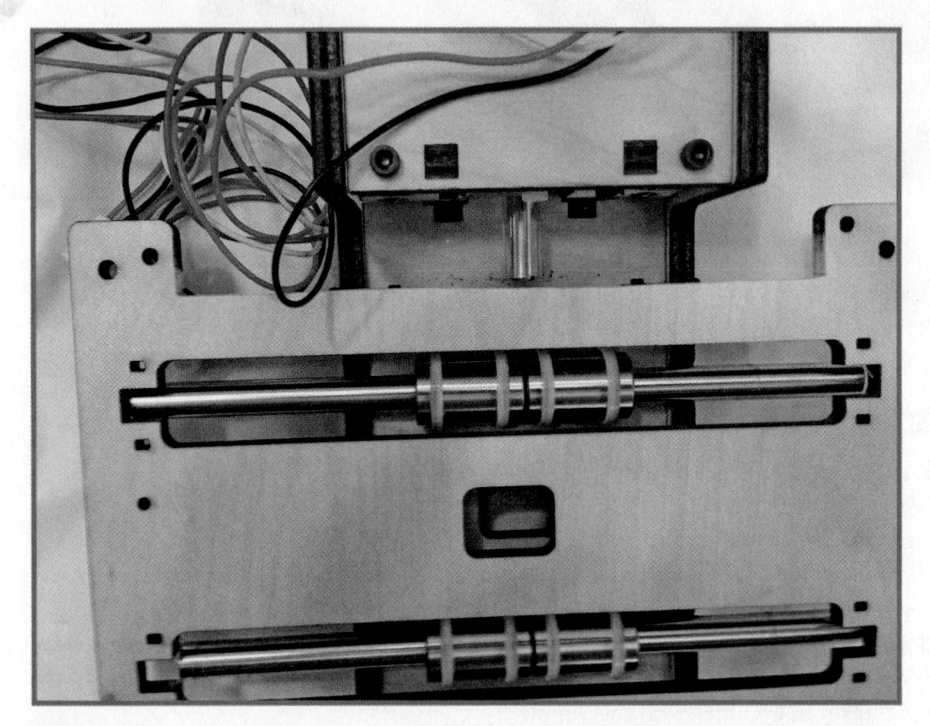

FIGURE 3.11 The bottom X-axis plate needs zip ties to lock it in place.

After you're satisfied that the bottom X-axis plate moves smoothly left and right (and the fishing line stays secure on the vinyl tubing), attach the print bed using the 3M bolts with the small springs between the top plate (print bed) and the bottom plate. By tightening these four bolts, you'll be able to level the print bed when you begin testing your printer.

It's critical that the print bed be completely level so that the tip of the hot end that is extruding hot plastic doesn't scrape against any portion of the wood print bed.

When you're done with the X axis, it's time to tackle the Y axis. The Y-axis steps are not complicated, but there are a lot of them! The Y-axis assembly starts when you attach four bearings on one side of a laser-cut wood piece and four bearings on the other side. (The step number may change, but this should help you know when you've reached the Y-axis point in the assembly.)

Again, pay attention to the location of the zip tie nubs as well as the orientation of the wooden pieces. Use the photos as visual clues. For example, in Figure 3.12, you'll see that an angled notch is cut into this normally square piece. That angled notch will help you orient the piece and show you where the zip tie nubs are to be located so they don't interfere with any moving parts.

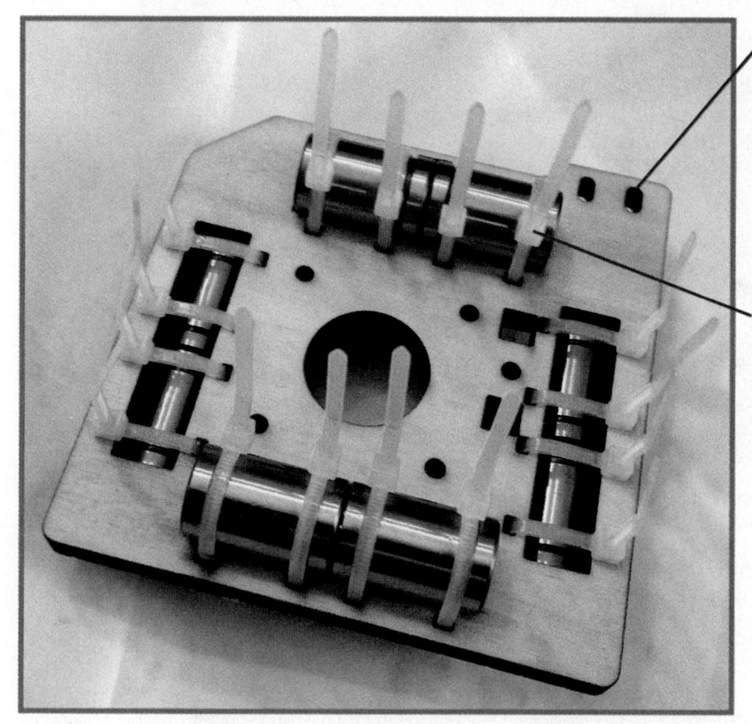

FIGURE 3.12 The Y-axis assembly begins with this piece.

Finish attaching a few more pieces to the wooden piece with all the bearings, and then bolt in the Y-axis motor (again, using photos to help you determine which side of the wooden piece the motor should be attached to as well as the direction of the wires). I'm going to call this Y-axis Mini-Assembly 1. Set this piece (with the motor) aside for a moment.

You'll now build the Y-axis Mini-Assembly 2, which is fairly straightforward. When you're done, you'll have a long Y-axis assembly that looks like the one in Figure 3.13.

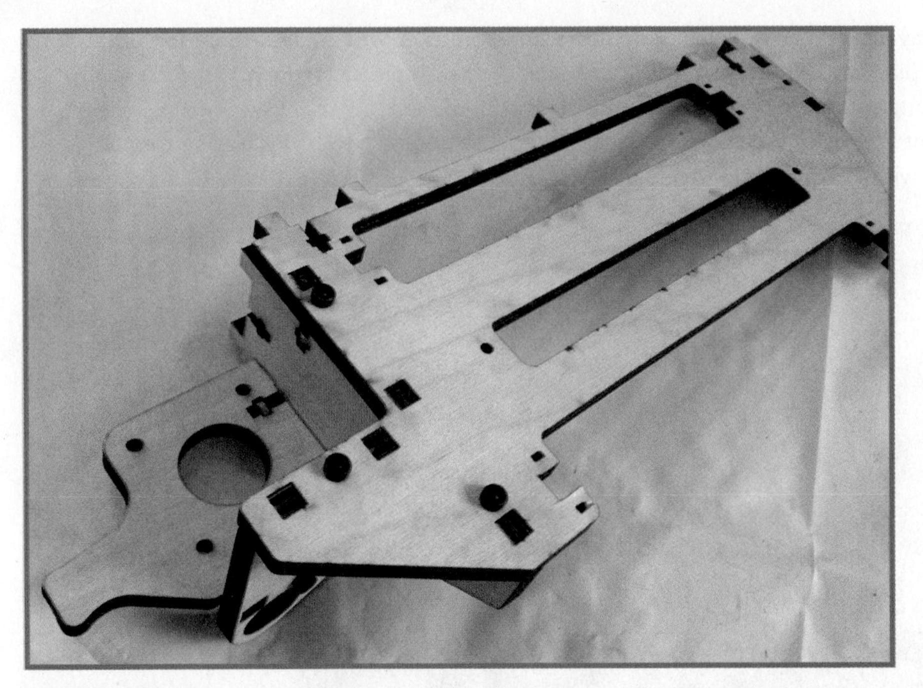

FIGURE 3.13 This assembly allows the hot end to move backward and forward.

Set the Y-axis Mini-Assembly 2 aside with Mini-Assembly 1, and next you'll tackle part of the Z axis. The Z axis differs from the X and Y axes in that it won't be using fishing line. Instead, the Z axis requires the use of the 10" lead screw with a piece of vinyl tubing zip tied to it, as shown in Figure 3.14.

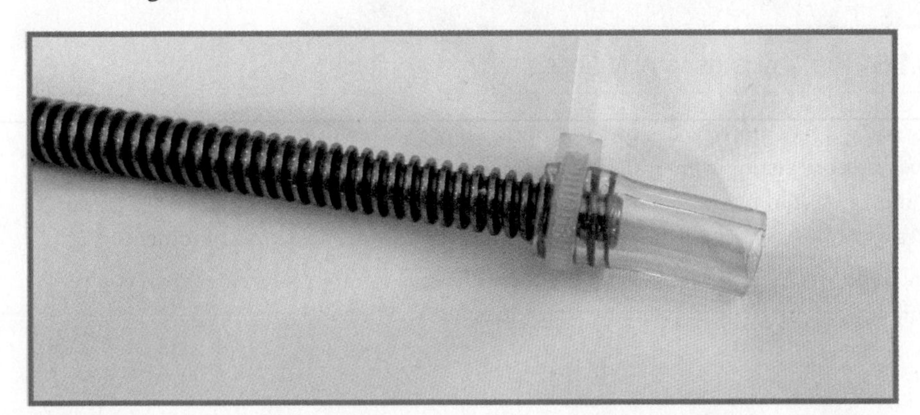

FIGURE 3.14 Use a zip tie to hold the vinyl tubing to the lead screw.

You then insert the open end of the vinyl tubing over the Z-axis motor and use another zip tie to hold it in place. Get that zip tie good and tight, as shown in Figure 3.15. Spin the lead screw with your hand; if the motor axis turns easily, cut the tail on the zip tie. Otherwise, tighten the zip tie a bit more.

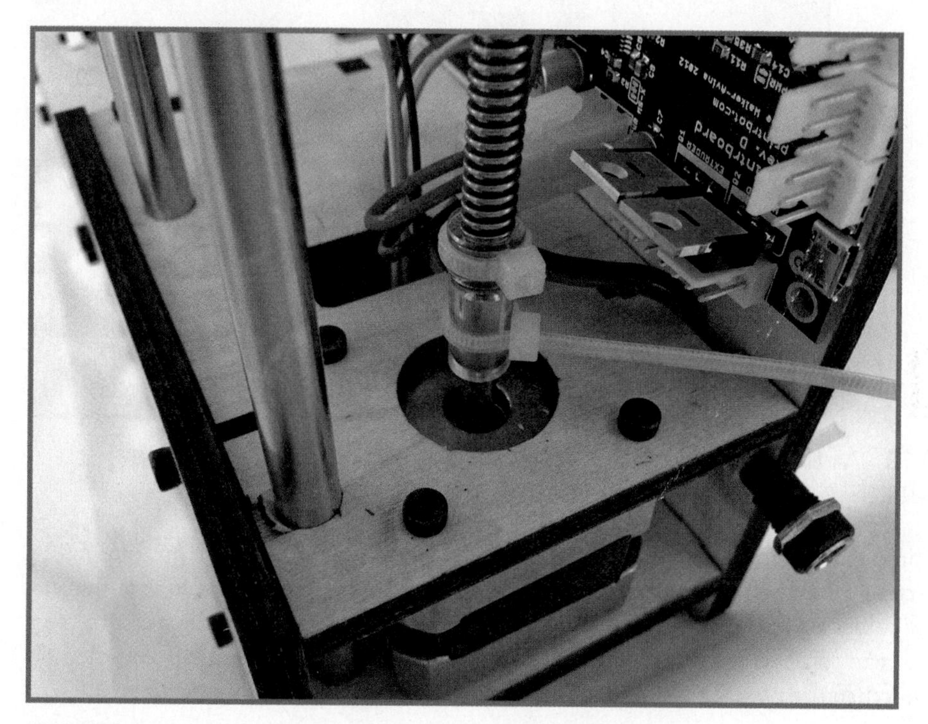

FIGURE 3.15 Attach the lead screw to the Z-axis motor.

End of Assembly Observations

Next, you'll tackle attaching the circuit board. This is easy, but be careful not to damage the circuit board or touch any of the sensitive electronics parts on its surface. It will help you to connect the power plug to the circuit board after you've zip tied the board to the side of the Simple, as shown in Figure 3.16.

FIGURE 3.16 All the wires from the motors will eventually connect to the motherboard.

Now it's time to attach both Y-axis mini-assemblies (1 and 2). Find the small nut that fits the lead screw. Slide the Y-axis Mini-Assembly 1 over the two 10" metal rods so that the lead screw comes up and through the cutout in the shape of the nut. Stick the nut into the hole and thread the lead screw into it. Make sure the nut stays in the (nut-shaped) hole, because you then use two M3 bolts to lock that nut into place, as shown in Figure 3.17.

FIGURE 3.17 The lead screw will control the up/down movement of the hot-end.

Next you'll insert two additional 6.5" metal rods through the Y-axis bearings. The same as with the X axis, you take the Y-axis Mini-Assembly 2 (shown back in Figure 3.13) and attach it to the Y-axis Mini-Assembly 1 with zip ties around the metal rods. Make certain these zip ties are as tight as you can get them - the tighter the better. And while you're at it, go ahead and tie a small knot and secure it with an M3 bolt and nut to the inside of the long Y-axis piece, as shown in Figure 3.18. (In Figure 3.18, you can also see the zip ties holding the longer Y-axis assembly.)

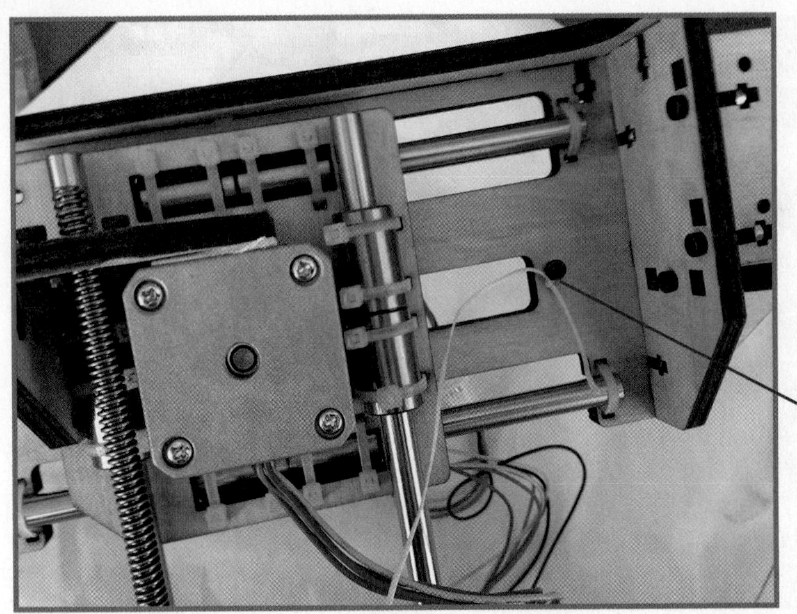

FIGURE 3.18 The Y axis assembled, and fishing line attached on one side.

Wrap the fishing line three to four times around the vinyl tubing and secure the other end with an M3 bolt and nut, keeping the line as tight as possible.

NOTE

I found that wrapping loosely and using my thumb to cover the end of the vinyl tubing prevented the fishing line from slipping off. After I had three to four wraps, I used my fingernail to push the fishing line together (making sure it didn't overlap on itself) while pulling on the free end and keeping tension on the line. If you've got someone to help, the person can hold and keep tension on the free end of the line as you wrap. Whichever way you do it, try to keep tension on the line as you wrap and secure it.

In the remaining steps, you assemble the extruder and hot-end assembly. You start by stacking a mix of five odd-shaped parts and securing it all with three M3 bolts as shown in Figure 3.19.

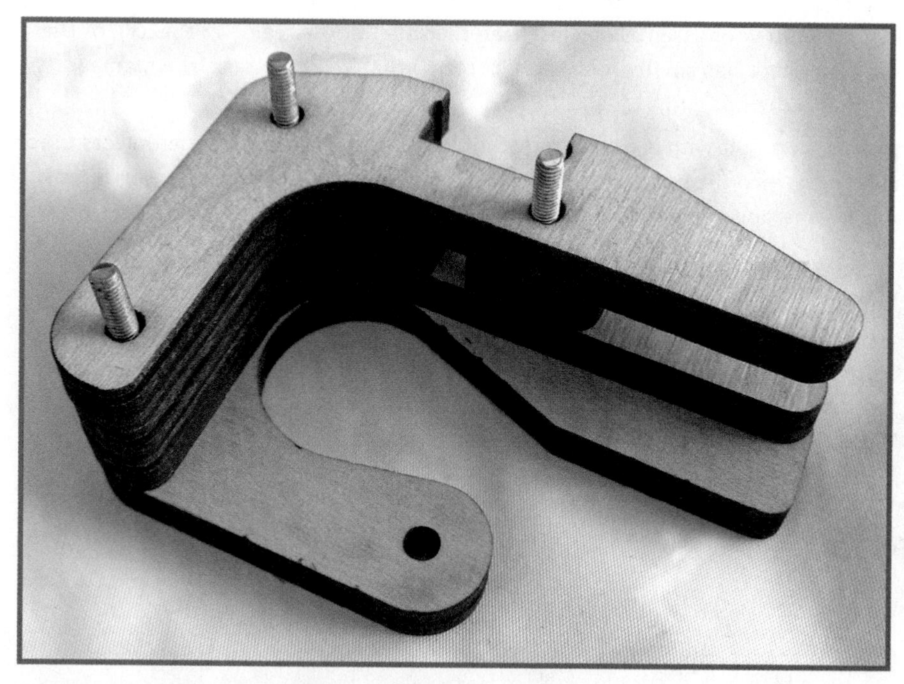

FIGURE 3.19 Starting the extruder assembly.

In additional steps you assemble smaller bits (some using wood glue) that end up looking like the item you see in Figure 3.20. This little collection of parts puts tension on the plastic filament as it is being fed into the hot end, so go slowly, making sure all the pieces and parts are oriented properly, and attach it carefully!

FIGURE 3.20 More bits that make up the extruder assembly.

You begin wrapping up your assembly by inserting the small driver gear over the axle of the extruder motor. Notice it has small grooves on it (teeth) that will grab onto the filament and help feed it into the hot end.

Insert the hot end, secure it with a couple of M3 bolts, add the last few wooden pieces used to direct the filament into the hot end, and you're done! When you've finished your Simple, it should look like the one in Figure 3.21.

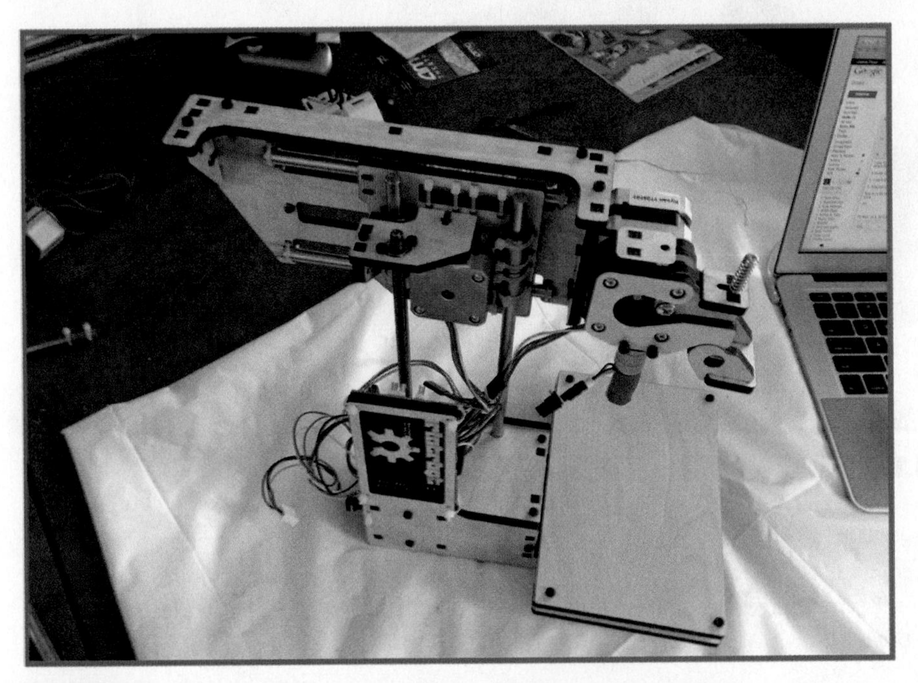

FIGURE 3.21 A finished Printrbot Simple, ready for testing.

Connecting All Wires

All that's left now is to connect up all the various wires to the circuit board. You'll want to look carefully at the circuit board and attach the wires to their respective plug-in ports. For example, the X-motor's wires all terminate in a white plug that looks like the one in Figure 3.22.

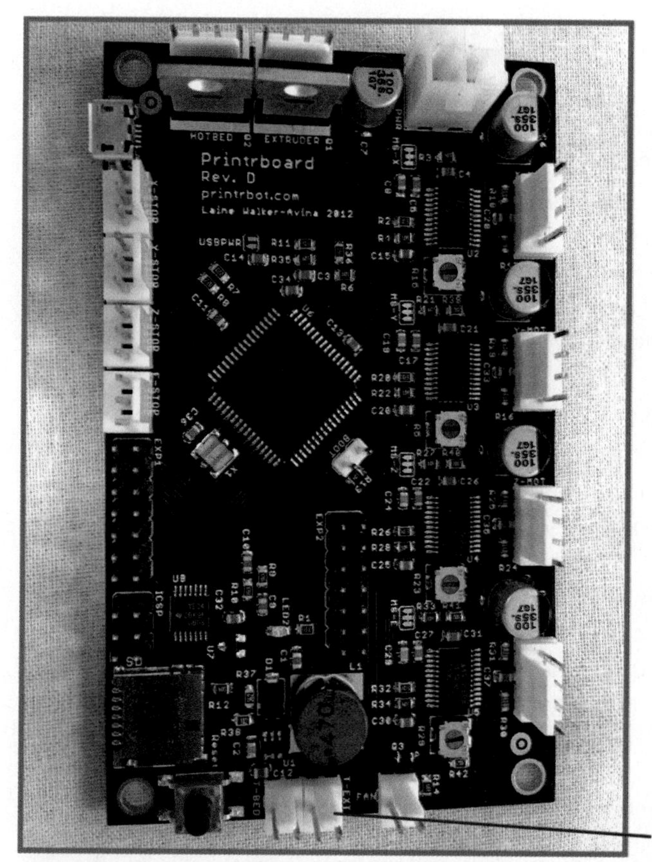

FIGURE 3.22 A closeup of the Simple circuit board.

I'm hesitant to point out any particular ports on the circuit board because these could easily change in future versions, but one thing I do wish to point out is that every port is labeled. This makes it easy to match up the motors and hot-end to their proper ports. For example, in Figure 3.22, if you look at the very bottom of the board, you'll see a white port with the letters T-EXT. The hot-end is constantly sending its current temperature to the circuit board (and on to the software used in 3D printing), and you'll need to make sure to connect the proper wire coming from the hot-end to this port.

Likewise, you'll find motor connectors running down the left side of the circuit board. In Figure 3.22, motor ports on the circuit board have a label like Z-MOT, Y-MOT, X-MOT, and E-MOT. These correspond to the four motors, Z, Y, X, and Extruder, respectively.

Again, you'll want to consult the Getting Started Guide for complete instructions on wiring up your own Simple.

Finishing Thoughts

My total build time was less than four hours. Keep in mind that I was shooting photos and taking notes, so I might have been able to complete it in three hours or less. But I wasn't in a hurry, and you shouldn't be either.

There's a reason this 3D printer is called the Printrbot Simple. It's not hard to build. You have to take your time, examine the photos, read the assembly instructions, and look over all the parts in your kit to make sure you've got everything. It's amazing that the only tools required were two Allen wrenches (2.5mm and 1.5mm), a Phillips-head screwdriver, pliers, scissors, wood glue, and my two hands. I've put together furniture that required more tools than this!

Still, I know that the first time building a 3D printer can be a bit stressful. I recommend frequent breaks. If you hit a snag or get frustrated, walk away for a few minutes.

Sometimes, skipping ahead in the instructions can also help. I don't mean you should jump ahead and start building, but simply observe. Look at photos and read instructions, and you'll find that sometimes a question you have is answered a few steps forward. I found myself doing this a few times when I was trying to understand how one or two small pieces were oriented with respect to other pieces. Another time, I jumped ahead to see a photo that showed me the proper orientation of the wiring on one motor. Take advantage of the dozens of full-color photos that often help you out.

NOTE

Don't ignore the Printrbot forum! You'll find that each of the Printrbot models, including the Simple, has its own area on the forum, and plenty of "experts" lurk there who are happy to help someone new get a 3D printer working and printing.

Keep in mind that the assembly instructions for the Simple are a work in progress. If an error is found, it's usually fixed quickly and an update is posted online. In fact, as I wrote this, I went over the assembly instructions and found two new steps for creating a flat surface on the extruder motor's axle with a file. Those weren't there yesterday, and I'm glad I found them because it's an easy thing to remove the extruder motor and perform this step.

So, I'll wrap up this chapter by summarizing some of the lessons I learned during my Simple build:

- Don't overtighten your M3 bolts.
- Look at photos to determine where a zip-tie nub is placed (and on what side of the wood).
- An extra set of hands is always nice, especially when wrapping the fishing line around the vinyl tubing on the X and Y motors.

- Use a ruler to measure the length of bolts; don't eyeball them.
- If a step doesn't call for wood glue, don't use it!
- Use a screwdriver and pliers to tighten a zip tie one or two more clicks.
- When hammering in the 10" metal rods for the Z axis, use a rubber mallet and go slowly. The metal rods do *not* need to go all the way through the holes on the very bottom of the Simple; 3/4 is fine.
- When in doubt, contact tech support at (http://www.jotformpro.com/Printrbot/support-ticket) or post a question on the forum (http://www.printrbottalk.com/forum/)

I wish you the best of luck as you build your Simple! It was fun to build, and I cannot wait to fire it up and start printing, which is the focus of Chapter 4, "Configuring the Software." I also show you how to test your Simple to make sure everything works as designed, and then we print a test item. Let's go!

Configuring the Software

Successfully building a 3D printer is an amazing accomplishment. If you've finished building your own Printrbot Simple (or have just received a preassembled unit), you're ready to move forward with the next step of the 3D printing process. It's all about the software now, and as you'll discover in this chapter, you need to know a few things about the specialized software used by hobbyists to print in plastic.

If you're ready to get started with the software part of the process, you should download and read over the "Simple Getting Started Guide" that the Printrbot team has created just for the Simple. It's a PDF file that explains which software to download and install and how to test your Simple to make sure everything is working properly. I cannot stress enough that it's important to read through the "Simple Getting Started Guide" completely before you download and install the software. To grab the guide, open a web browser and point it to the following web address, as shown in Figure 4.1:

http://printrbot.com/support/instructions-and-guides/

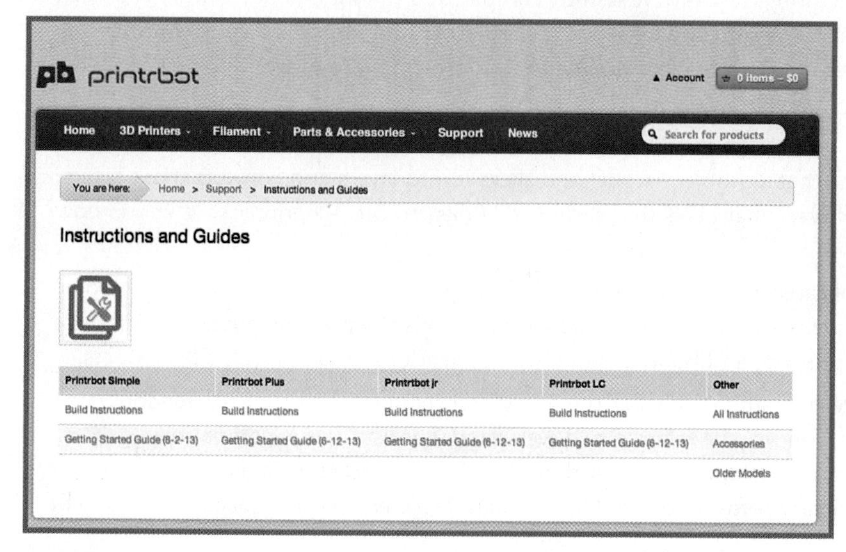

FIGURE 4.1 The Download page for the "Getting Started Guide."

Types of 3D Printing Software

I promised you early in the book that I would keep the technical discussions as easy to follow as possible, so first I want to reduce as many of the acronyms and strange terminology as possible. You'll pick up on this stuff anyway, after you get more confident in using your 3D printer, but when you're just beginning, this is what you need to understand:

- You'll use special software to create 3D objects. This software is sometimes called CAD software; CAD stands for Computer Aided Design. It is software that helps you do the complicated stuff, such as creating straight edges, curves, and ensuring dimensions are correct (such as the length of the side of a cube) and all the other tasks involved in taking an idea in your head and making it appear on the screen of your computer.
- You also use special software to slice up that 3D object into layers—remember layers from Chapter 1, "The Big Question—What is a 3D Printer?" Your 3D object will be printed one layer at a time, with molten plastic coming out of the hot end. A special bit of software is needed to cut that 3D object into the various layers that, when stacked, make up the physical 3D object you'll be able to hold in your hands. This software takes each slice and converts it into special data (called g-code) that tells the hot end where to move as it lays down a bead of molten plastic.
- Another type of special software controls the electronics in your 3D printer. The three motors that control movement along the X, Y, and Z axes must have instructions given to them that control the direction that the motor axles rotate. The hot end must be instructed to heat up to a specific temperature that starts melting the plastic. And there's much more. This controlling software is sometimes referred to as Computer Aided Manufacturing, or CAM. It sounds complicated (and it can be), but all you need to know is that CAM software takes the g-code generated for each slice of your 3D object and uses that code to instruct the axes' motors how to move.

I've simplified the software explanation greatly, but honestly, it doesn't need to be complicated. In a nutshell you need software that lets you create a 3D object model, software that takes that 3D object model and slices it into layers and converts those layers into code, and software that takes that code and "talks" to the 3D printer so it knows how and where to move.

For me, the complicated part of the software is finding the software you can use with your particular 3D printer. Dozens of options are available for software that can slice up objects and "talk" to 3D printers, and hundreds of CAD applications let users create 3D objects.

The simplest route is to start with what the 3D printer manufacturer suggests for software. In some instances, you may be referred to a web page that lists all the compatible applications that will work with your brand of 3D printer. In other instances, some 3D printers use proprietary software created by the manufacturer, so the choice may already be made for you.

When it comes to the Printrbot Simple, you aren't locked in to any one type of software, but what's nice is that Printrbot suggests one particular application in the "Simple Getting Started Guide" and walks you through all the configuration settings that are specific to the Simple. So that's what I'm going to do for this chapter; I'm going to download Repetier and show you how I go about connecting my computer to the Simple, and testing and configuring it before I begin printing.

Downloading the Repetier Software

I use a MacBook Air, but if you use a PC, you'll be happy to know that the software recommended by Printrbot is available for both Windows and Linux. Most of the figures you'll be seeing in this chapter are from my Mac, but you'll find that the software looks similar in most, if not all, of the configuration screens. Again, if you hit a snag, contact Printrbot or Repetier with your question or post it on a forum.

NOTE

The Repetier software is free to use, but if you really like it and enjoy it, you might want to consider making a donation to the company. You can visit http://www.repetier.com/donate-or-support/ to make a donation. The amount is up to you, but keep in mind that financial support means updates for the software and continued technical support.

I've downloaded the software to my Mac and double-clicked the icon to install it. If you're a Mac owner, you'll probably find that the software is from an "unidentified developer" and cannot be installed. Don't worry—the software is fine if you've downloaded it directly from Repetier. All you need to do is hold down the Control key and click the installation icon (and your login user account must be of the admin type that allows for installing software). A menu appears, and you should click the Open option at the very top. You'll see a window appear like the one in Figure 4.2. Click the Open button to run the Repetier application.

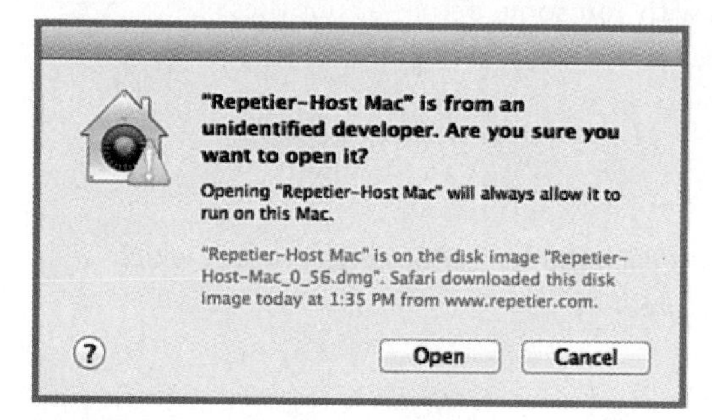

FIGURE 4.2 Installing the Repetier software on a Mac.

When Repetier first opens, you'll most likely see two open windows like the ones in Figure 4.3. One is the Preferences window for the Repetier software, and the other is the Repetier software user interface. Go ahead and close the Preferences window; in the rest of this chapter I focus on the settings you must make in the basic user interface.

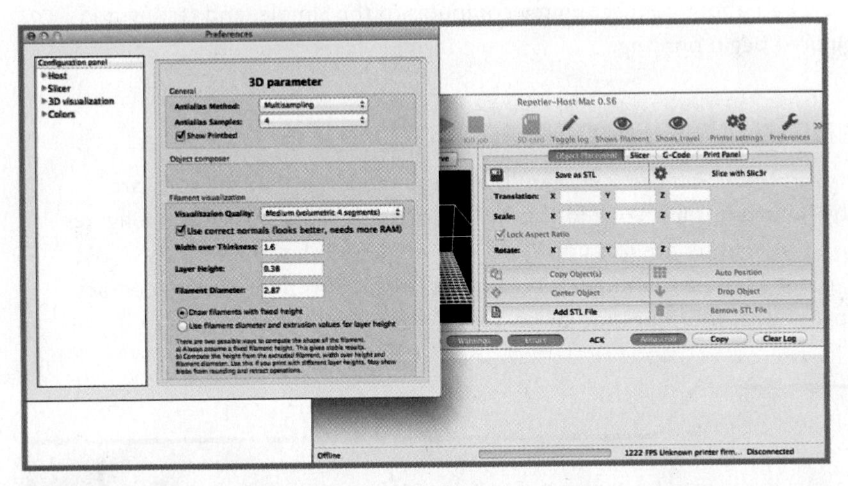

FIGURE 4.3 The Repetier software and its Preferences window.

NOTE

There are dozens and dozens of items to configure in Repetier, and Printrbot's "Simple Getting Started Guide" for the Simple shows you the most important ones to configure. However, you may want to consult the official Repetier configuration instructions by visiting http://www.repetier.com/documentation/ and clicking the link that corresponds to your operating system.

Because there are so few configuration settings to make prior to printing with the Simple, I'm going to go over most of them and share with you some details about these settings so you'll understand a bit more about their values and when you might have to change some of them.

Repetier Settings

The first Repetier configuration starts when you click the Printer Settings button, which opens the Printer Setting window shown in Figure 4.4.

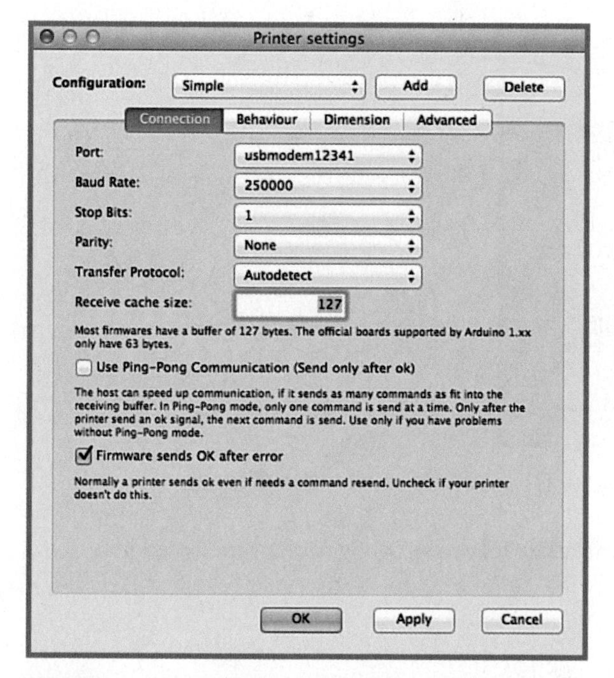

FIGURE 4.4 The Printer Settings window.

Click the Add button and a small window appears, like the one in Figure 4.5. Type in **Simple** and click Create.

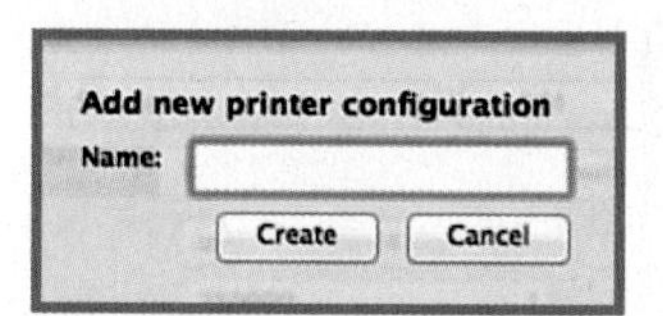

FIGURE 4.5 The Add window enables you to create a custom configuration.

Make certain you have the Connection tab selected on the Printer Settings window, and then make the following configurations as indicated in Figure 4.6.

1. Select the port from the drop-down menu that corresponds to the USB port you're using to connect the Simple to your computer.
2. Set the Baud Rate to 250,000 from the drop-down menu.
3. Change the Cache Size to 127.

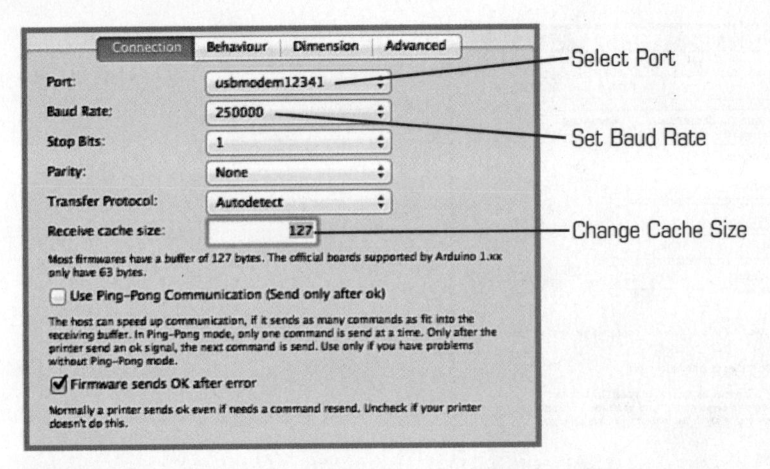

FIGURE 4.6 Connection settings allow Repetier to talk to the Simple.

Change to the Behavior tab next and make the following configurations indicated in Figure 4.7.

1. Change the Travel Feedrate to 500.
2. Set the Default Extruder Temperature to 195.
3. Set the Default Heated Bed Temperature to 0.
4. Make certain the three boxes are checked for Check Extruder and Heated Bed, Don't Log Temperature Requests, and Disable Extruder After Job/Job Kill.

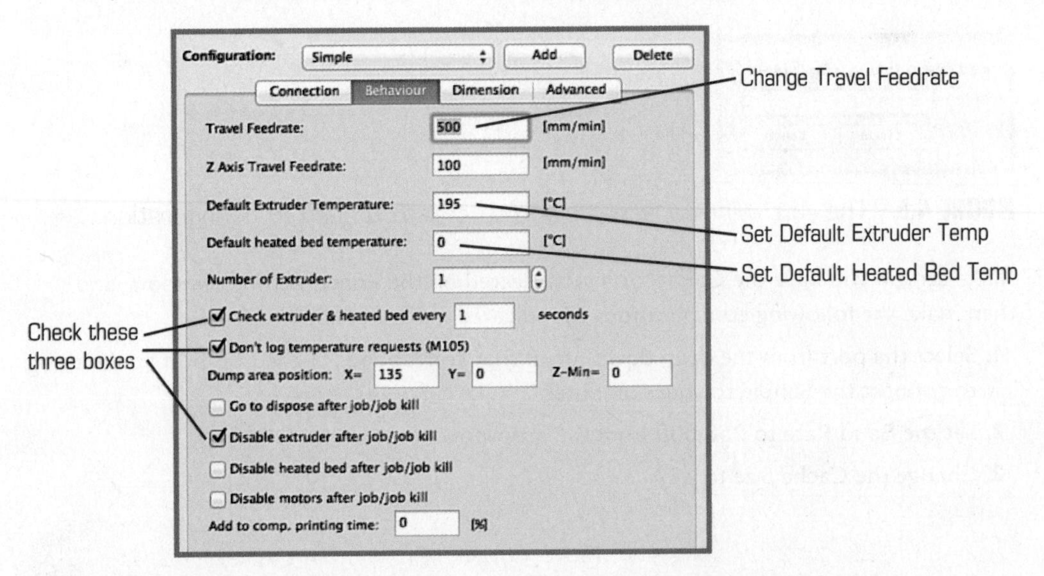

FIGURE 4.7 Temperature and travel speeds must be set for the Simple.

If you choose to later add a heated bed to your Simple, you should return to this setting and change the Default Heated Bed Temperature to the temperature recommended for that particular heated bed.

NOTE

A heated bed helps to ensure that the molten plastic cools at a slower rate, helping to prevent shrinking and warping. As of this writing, no heated bed is currently being sold by Printrbot for the Simple, but check the printrbot.com website to see if one is available if you are interested.

Switch to the Dimensions tab and set the configurations indicated in Figure 4.8.

1. Set the X Max and Y Max values to 100 (mm).
2. Set the Print Area Width to 100 (mm).
3. Set the Print Area Depth to 100 (mm).
4. Set the Print Area Height to 100 (mm).

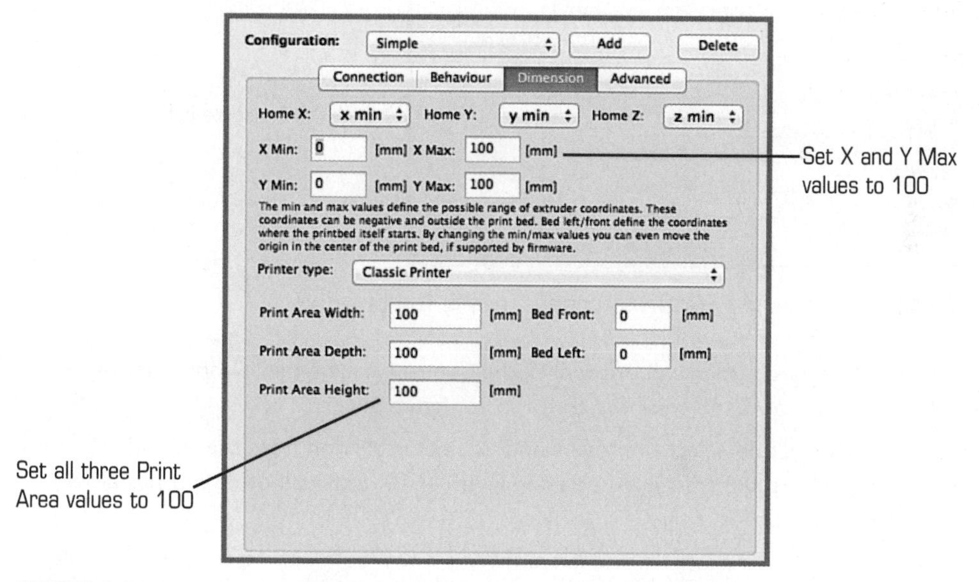

FIGURE 4.8 Define the dimensions of the Simple's print area.

These values (100mm) define the boundaries of the area where the Simple's hot end can deposit melted plastic. Picture a cube sitting on the print bed that's 100mm on each side, and you'll have a better idea of the maximum size for any object you want to print with the Simple. If an upgrade kit is ever offered by Printrbot that allows you to increase the width of the print bed (the X axis), this is where you'd go to change the value for the X Max and the Print Area Width.

Likewise, if longer metal rods and lead screws were made available for the Z axis, you could increase the maximum height of any object you could print.

When you're done with the Dimensions tab, click the Apply button and then click the OK button to save your configurations.

Slic3r

The Repetier software can also perform the slicing calculations required to take a 3D object, slice it up into layers, and then convert those layers to g-code. Again, the "Getting Started Guide" walks you through the proper settings for use with the Simple.

It all starts with clicking the Slic3r tab shown in Figure 4.9 and clicking the Configure button.

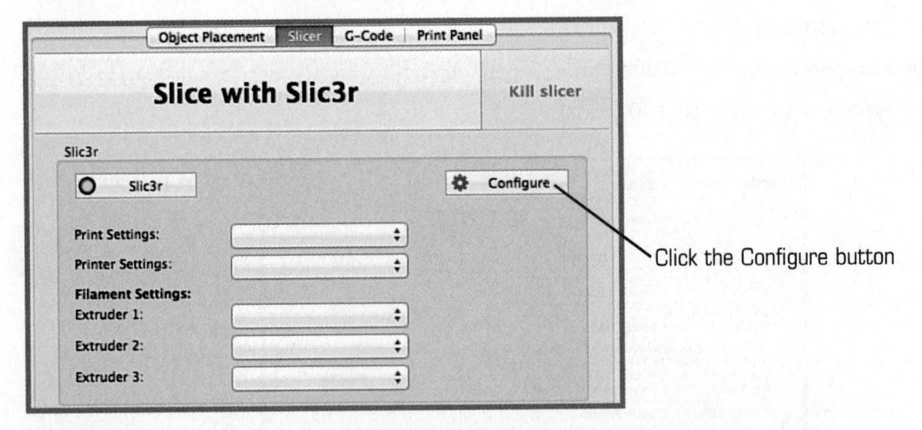

FIGURE 4.9 Configuring Slic3r to create layers for printing.

A configuration wizard screen will open; click the Cancel button—the "Simple Getting Started Guide" shows you all you need to do to configure Slic3r.

On the Print Settings tab, select the Layers and Perimeters option from the left-side list and set the Layer Height to 0.3 (mm), as shown in Figure 4.10. Leave all other settings alone, and click the Infill option.

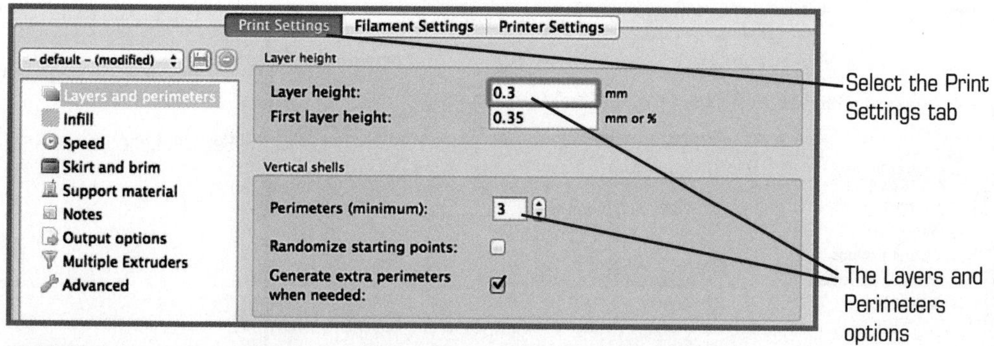

FIGURE 4.10 Repetier can control the height of print layers.

For the Fill Density, change the value to 0.3. Click the Fill Pattern drop-down menu and select Rectilinear from the various options, as shown in Figure 4.11. Make no other changes and then click the Skirt and Brim option.

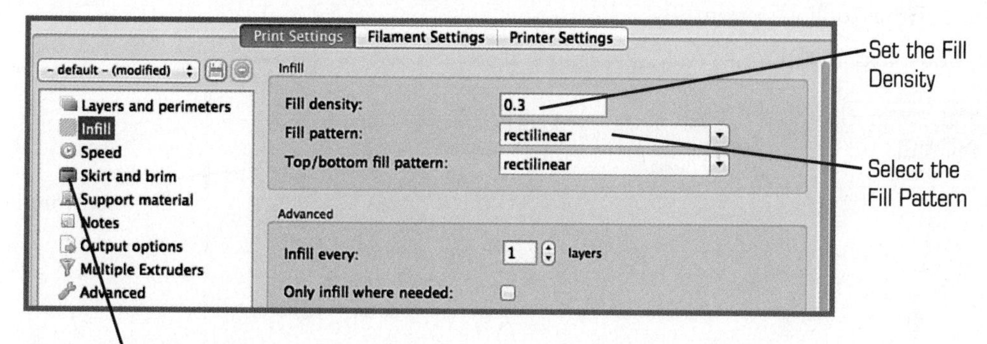

FIGURE 4.11 Configure the Infill details for a printed object.

On the Skirt and Brim screen, set the Loops option to a value of 2, as shown in Figure 4.12. (These will force a number of test loops to be printed around your object on the print bed before the actual printing of the object begins. It's a great way to make certain the molten plastic is coming out of the nozzle and adhering to the print bed.) Make no other changes and then click the Filament Settings tab.

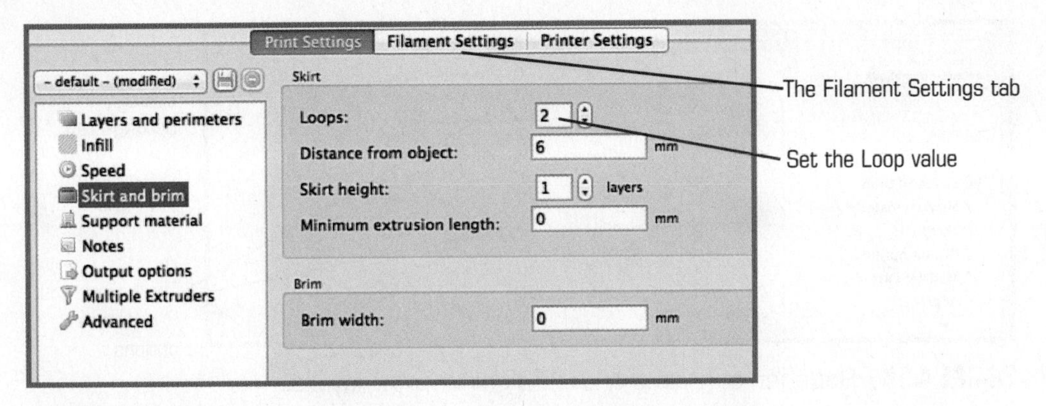

FIGURE 4.12 Create a test loop around your printed object.

Make sure the Filaments option is selected on the left side of the screen, and then change the Diameter value to 1.70 (mm), as shown in Figure 4.13. Make no other changes and then click the Cooling option.

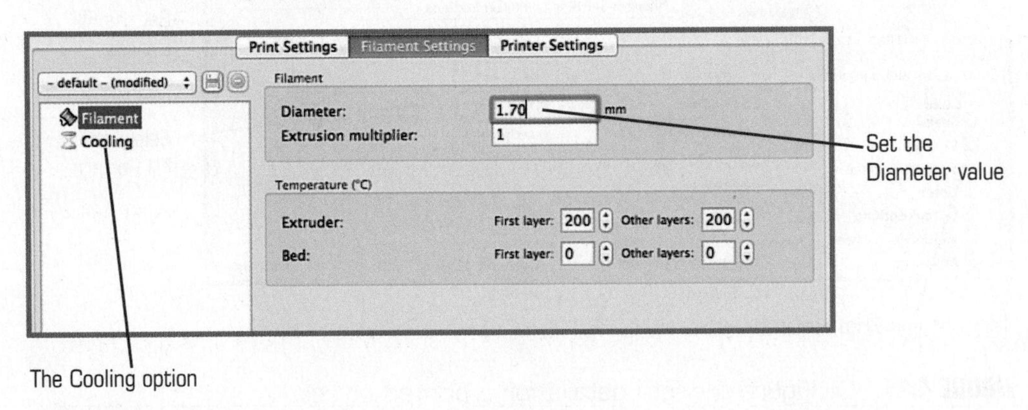

FIGURE 4.13 Set the filament diameter size used by your 3D printer.

The Simple has no fan, so you should uncheck the Enable Cooling box, as shown in Figure 4.14. Make no other changes and then click the Printer Settings tab to the right of the Filament Settings tab.

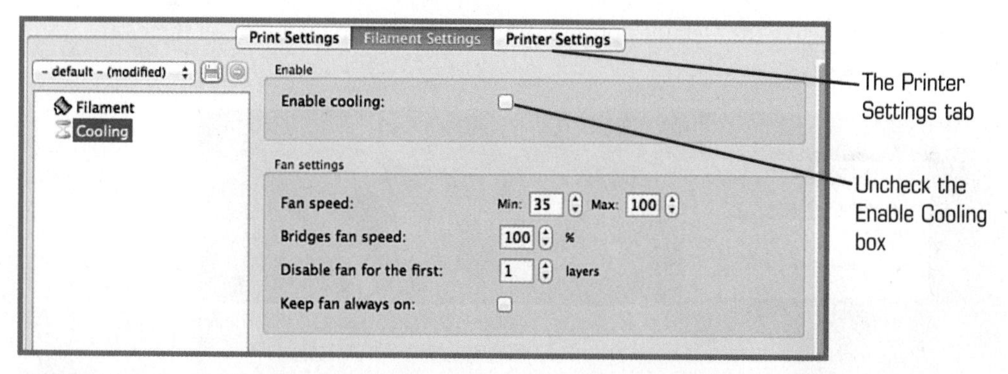

FIGURE 4.14 The Simple has no fan, so configure Repetier accordingly.

On the Printer Settings tab, select the General option on the left side of the screen and set the Bed Size X and Y values each to 100 (mm). Set the Print Center X and Y values each to 50 (mm), as shown in Figure 4.15. Make no other changes and then click the Extruder 1 option.

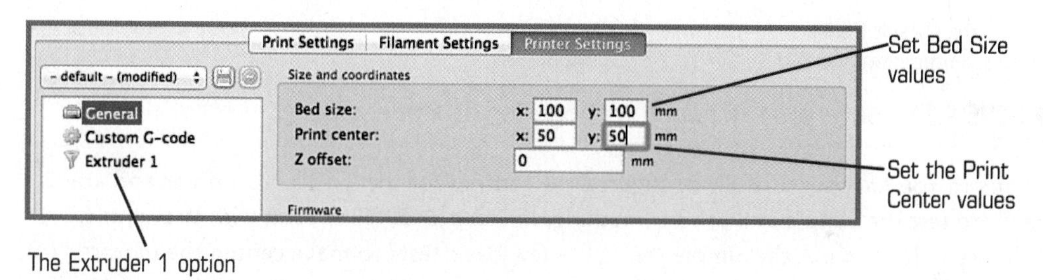

FIGURE 4.15 Configure settings related to the print area of your 3D printer.

On the Extruder 1 option page, change the Nozzle Diameter value to 0.4 (mm), as shown in Figure 4.16. Make no additional setting changes.

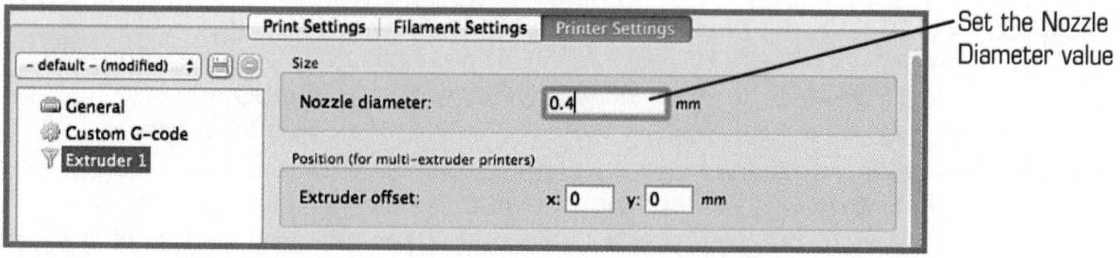

FIGURE 4.16 Configure Repetier with the nozzle diameter for your 3D printer.

You've made a handful of configuration settings on three tabs: Print Settings, Filament Settings, and Printer Settings. Before you begin using the Simple, you must save the changes as the default settings. To do this, go back to each tab and click the Save Current Printer

Settings button. Select the Default option from the drop-down menu for all three tabs, as shown in Figure 4.17.

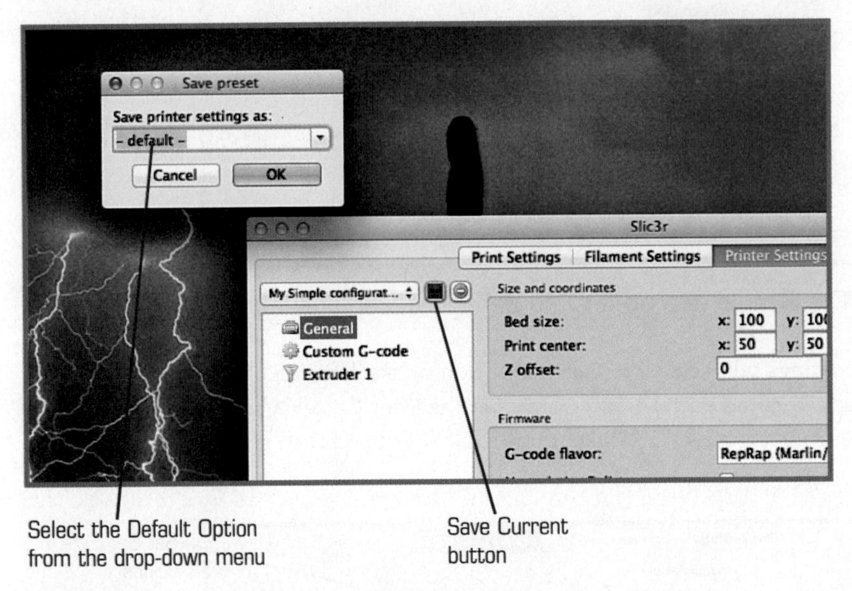

FIGURE 4.17 Save your Repetier settings for all three tabs.

That's it! You have successfully configured the settings for the Simple 3D printer and are ready to test the Simple and print something. In Chapter 5, "First Print with the Simple," I show you how to put the Simple through a few basic tests to make certain the motors are working and the hot end is reaching the correct temperature to melt the PLA plastic filament that will be fed in by the extruder.

NOTE

You've probably noticed a lot of strange terms are found in Repetier. Most of these items (such as Density or Skirt Height) can simply be configured based on Printrbot's recommendations, but if you're really wanting to delve deeper into 3D printing, you'll want to spend some time reading over the Repetier Help files for a better understanding of the terms and how they apply to your 3D printer. This is a beginner level book, and a complete understanding of all of these strange settings isn't required to use your 3D printer, but it will help as you gain skills and seek to print more advanced objects, so take some time to start your own research into all of the settings offered by Repetier (or the software you choose to control your 3D printer).

5

First Print with the Simple

Now that the Simple is assembled and the software is configured, there's only one thing left to do—let's print something!

In this chapter you'll be reading about the last few tasks required to print a 3D object model with the Simple using the Repetier software. Given that more than 100 different models of 3D printers are out there, this chapter can't possibly cover every situation. But keep in mind—even if you're not using Repetier, even if you're not using a Printrbot Simple—there is a fairly standard list of tasks that you must do to print an object on all 3D printers:

- You'll need to download or create an STL file on your computer.
- You'll need to import an STL file for printing into the software.
- You'll need to connect your 3D printer to your computer.
- You'll need to get the Hot-end up to proper temperature.
- You'll need to slice that object into layers.
- You'll need to "home" the nozzle end of your 3D printer's hot end (more on that in a moment).
- Print!

NOTE

STL file—this is a new term for you. STL stands for stereolithograpy, a form of 3D printing (that you'll read about in Chapter 10, "Alternatives to the Printrbot Simple") that used this special file format for storing data to print a 3D object. STL files are pretty common these days for all forms of 3D printing, including with the Printrbot Simple.

Not all of these tasks must be done in the order specified, but certain tasks (such as beginning the print job) won't start unless others have already been completed (such as slicing the object into layers).

In this chapter I walk you through the steps I do each time I want to print an object with my Simple. If you've got a different model of 3D printer, not everything will match up. That's when it's time to consult the documentation for your own 3D printer and figure out the list of tasks that required for you to print something. Keep in mind that there are lots of folks who have gone before you, so reach out to them in the forums if you need help. The amazing thing about the 3D printing hobby is how friendly and helpful the community is with novices; we all want this hobby to grow and prosper, and that means lending a hand to those who have questions.

So, without further ado, let me print something for you.

Downloading an STL

Before I can print anything with my 3D printer, I need to decide on what I'd like to print. I have designed a number of objects with Tinkercad (see Chapter 7, "Creating a 3D Model with Tinkercad," for a walkthrough of a 3D object being created from scratch), and I can certainly choose one of them for my first print job on the Simple. Lots of 3D printer owners choose to print what are called "calibration" objects—they allow users to print out an object and compare the accuracy of angles and widths and lengths to the actual 3D model to determine if the printer is working properly. But I want to jump right in and print something of mine. If I discover my printer needs some calibration, I'll do that a bit later. Right now, however, I just want to print something fun.

NOTE

Printrbot's "Simple Getting Started Guide" recommends a specific 3D model from Thingiverse called the 5mm Calibration Cube with Steps; you can view (and download) that model by visiting the following link: http://www.thingiverse.com/thing:24238. If you don't have the STL file for a 3D object to print, grab this one and perform the same steps I do in this chapter for my medallion object.

Figure 5.1 shows the Hello World file that I created in Tinkercad (see Chapter 7) and saved to my computer.

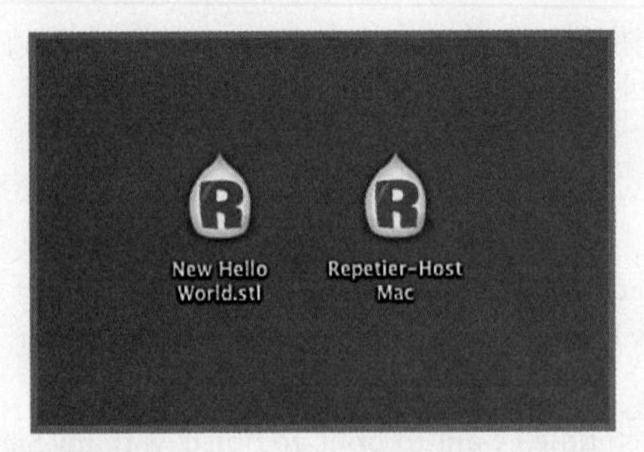

FIGURE 5.1 A 3D model's STL file.

Notice that its icon is a copy of the Repetier software icon. If I double-click it, the Hello World STL file opens in Repetier, as shown in Figure 5.2.

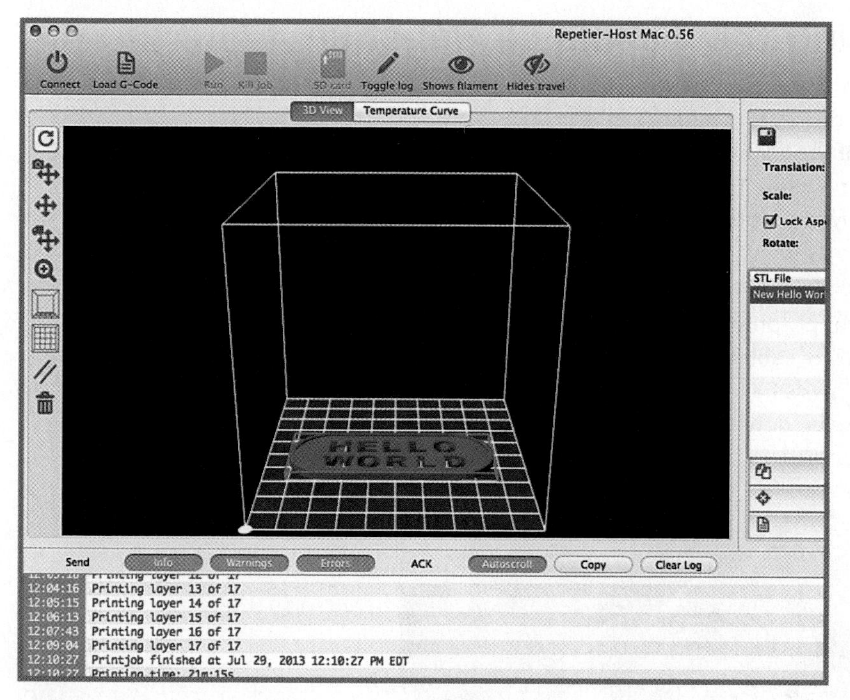

FIGURE 5.2 The STL file as viewed in Repetier.

This object is a small, thin, rectangular plate (with rounded ends) that has the words "Hello World" created via holes in the solid object's body. Right now, I'm not concerned about all the buttons and values being displayed in Repetier. Instead, I want to open up the STL file and confirm that I can view it in the 3D Viewer. (Click the 3D View button if you don't see your STL file after double-clicking it.)

NOTE

Before you begin printing, it is recommended that you put down a strip or two of blue painter's tape onto the print bed. When the plastic cools, it will adhere quite strongly to the wooden print bed surface—the painter's tape will allow you to more easily remove a cooled, printed object because the plastic doesn't stick as securely to the tape's surface as it does with wood.

Connecting the Simple to Repetier

Now that I've imported my STL file into Repetier, it's time to connect my 3D printer to my computer. I do this by first plugging in the USB cable to the Simple's circuit board and the other end of the USB cable into my computer. I already configured Repetier back in Chapter 4, "Configuring the Software." so that my computer knows which USB port to use. (Fortunately, my laptop has only one USB port, so Repetier has it easy. If you've got multiple USB ports, try to remember to always connect the 3D printer to the same USB port on your computer.)

Next, I need to apply power to the Simple using the AC adapter that came with the kit. I first plug the power cable into the wall and then insert the other end into the power port on the circuit board. A small green LED lights up on the circuit board to indicate the Simple is receiving power, as shown in Figure 5.3.

FIGURE 5.3 The green LED indicates all systems are go!

Now it's time to test to see if Repetier can communicate with the Simple; to do this, all I have to do is click the red Connect button in the upper-left corner of Repetier, as indicated in Figure 5.4.

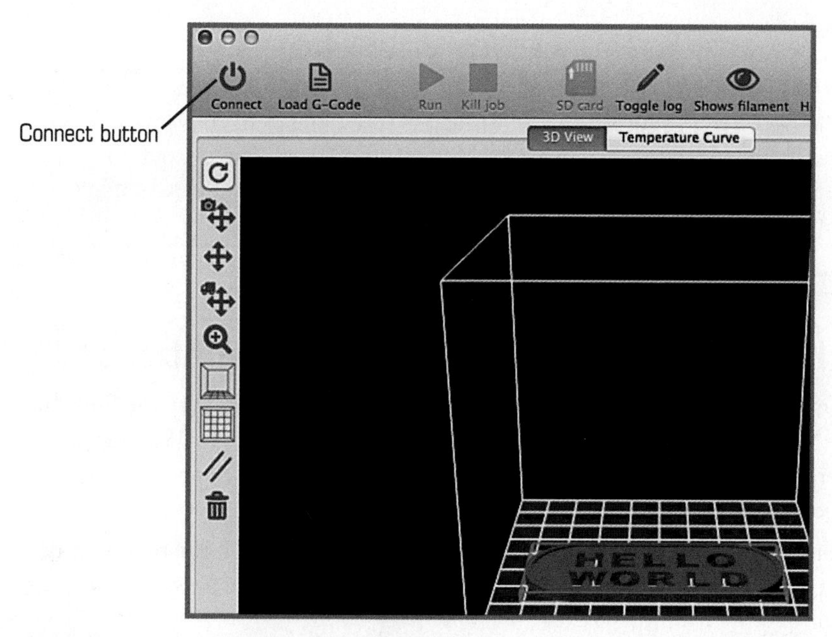

FIGURE 5.4 Click the Connect button.

If everything is connected properly, I should see the Connect button change from red to green. I also get a confirmation of the connection from the text that appears at the bottom of the screen, as shown in Figure 5.5.

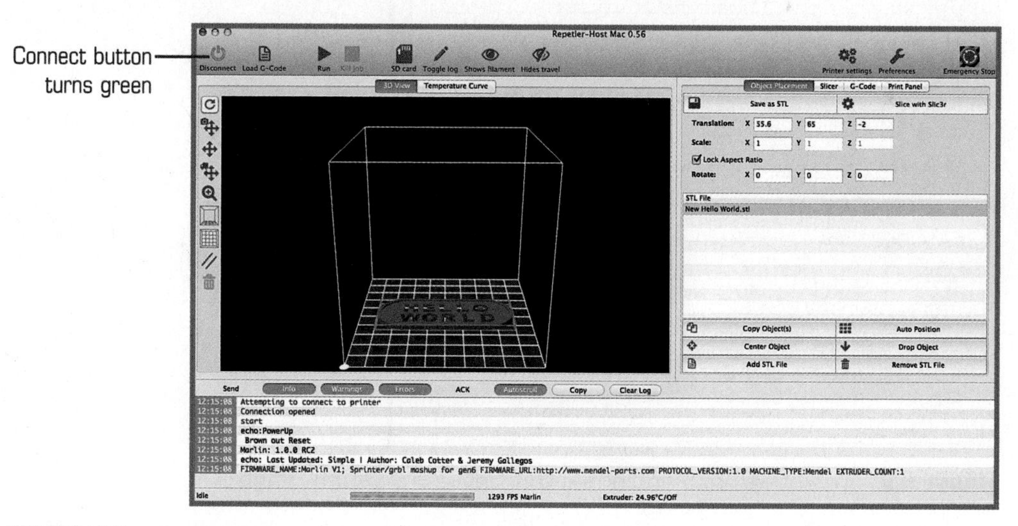

FIGURE 5.5 Confirm that your 3D printer is connected to Repetier.

NOTE

Consult the "Simple Getting Started Guide" at http://printrbot.com/support/instructions-and-guides/ for troubleshooting help with the Simple and Repetier software. You might also want to visit printrbottalk.com and post a question if you can't find an answer to your technical issue.

Get the Hot End Up to Proper Temperature

The hot end must get up to the proper temperature for melting the filament. The Printrbot Simple "Getting Started Guide" will have you configure the temperature to 195 degrees Fahrenheit (refer back to Chapter 4), and that's the temperature you will see used for this chapter's print job.

To get the hot end heating up, change to the Print Panel in Repetier. On the right side of the screen in Repetier, I click the Print Panel button shown in Figure 5.6 and then click the Heat On button.

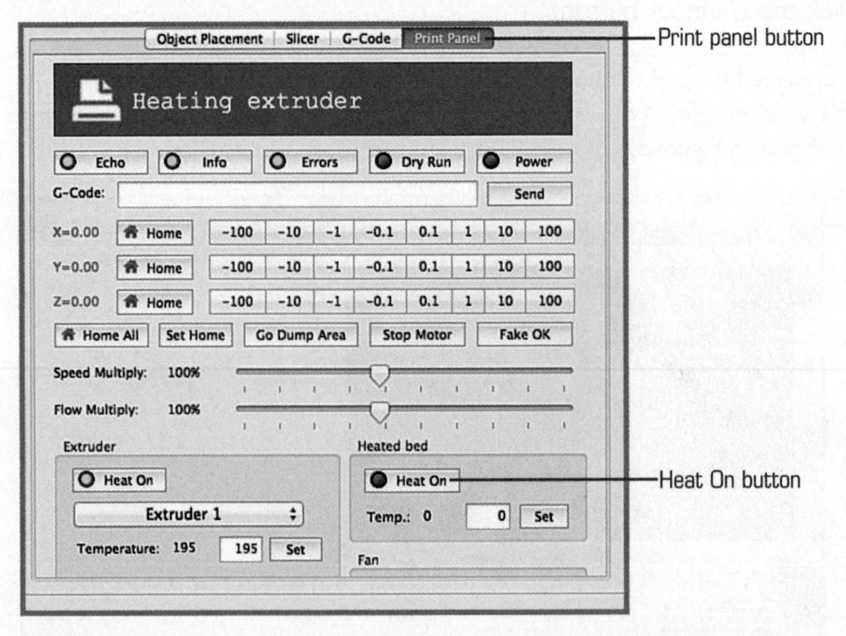

FIGURE 5.6 Turn the heat up to get the hot end melting plastic.

To verify that the temperature is climbing, tap the Temperature Curve button indicated in Figure 5.7, and take a moment to watch as the line climbs upward toward 195.

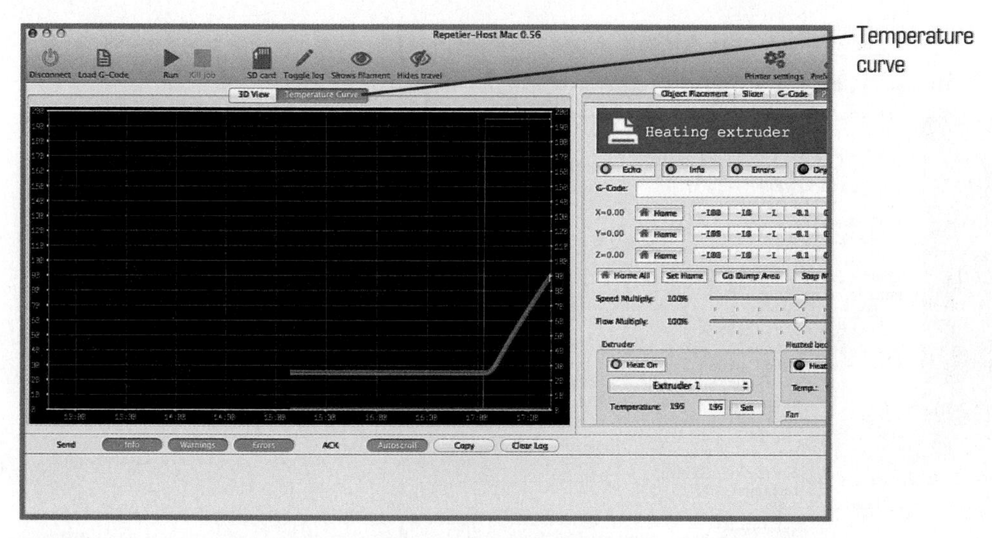

FIGURE 5.7 The hot end temperature increases.

While the hot end's temperature is increasing, be very careful not to touch it. You also want to make certain that nothing is touching the hot end, such as paper, wires, and so on.

Slice Your Object into Layers

Now it's time to slice up the object you've selected to print into layers. Click the Object Placement button indicated in Figure 5.8 and then click the Slice with Slic3r button.

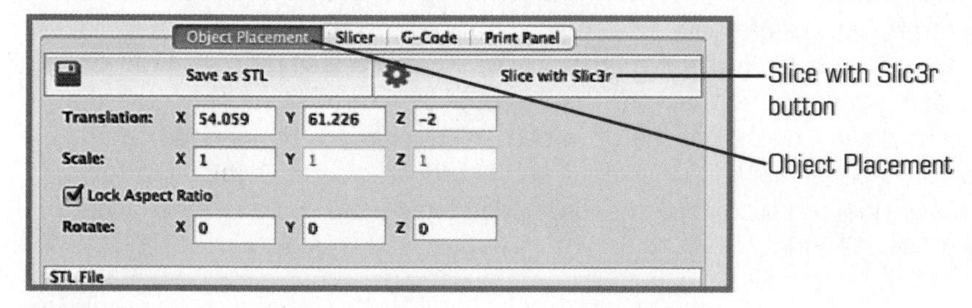

FIGURE 5.8 Use Slic3r to cut up your 3D object for printing.

After clicking the Slice with Slic3r button, you'll see some text appear in the window below the Object Placement button, as shown in Figure 5.9. This is the g-code that will be used to control the motors and Extruder, telling the motors how to rotate and in what direction.

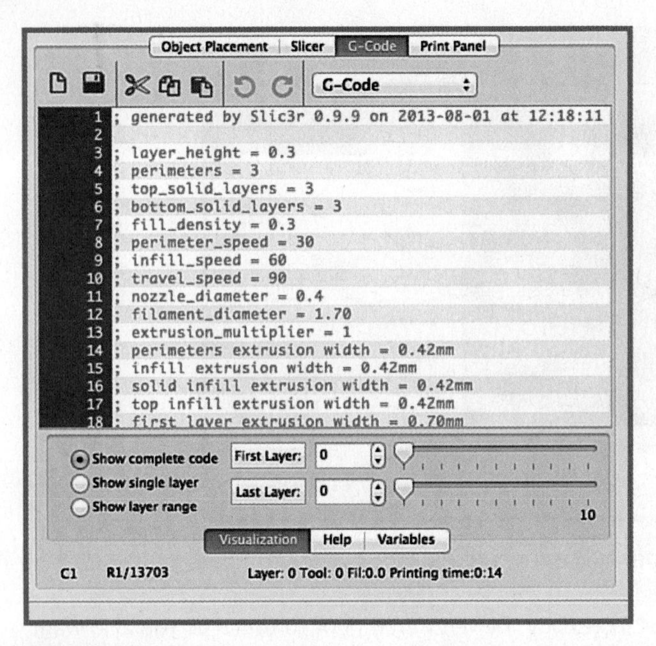

FIGURE 5.9 G-code for the 3D object.

NOTE

I mentioned g-code earlier in the book, but now you can see what it actually looks like. It's very cryptic stuff, but it's basically just code that tells the various motors how to operate. X10, for example, tells the X motor to spin in such a way that the print bed moves 10mm. The g-code Z-15 would tell the Z motor to spin the lead screw so that the hot end moves down (toward) to the print bed. You'll also see the various settings you configured for Repetier back in Chapter 4 submitted via g-code. If you'd like to see just how substantial the list of commands/codes for g-code really is, visit http://en.wikipedia.org/wiki/G-code.

Home the Hot End

Look at your Simple (or whatever 3D printer you are currently using) and take note of where the end of the hot end's nozzle is located. Figure 5.10 shows where my Simple's nozzle is located. It's about 5cm above the print bed and slightly to the left of dead center. It's also more toward the rear of the print bed.

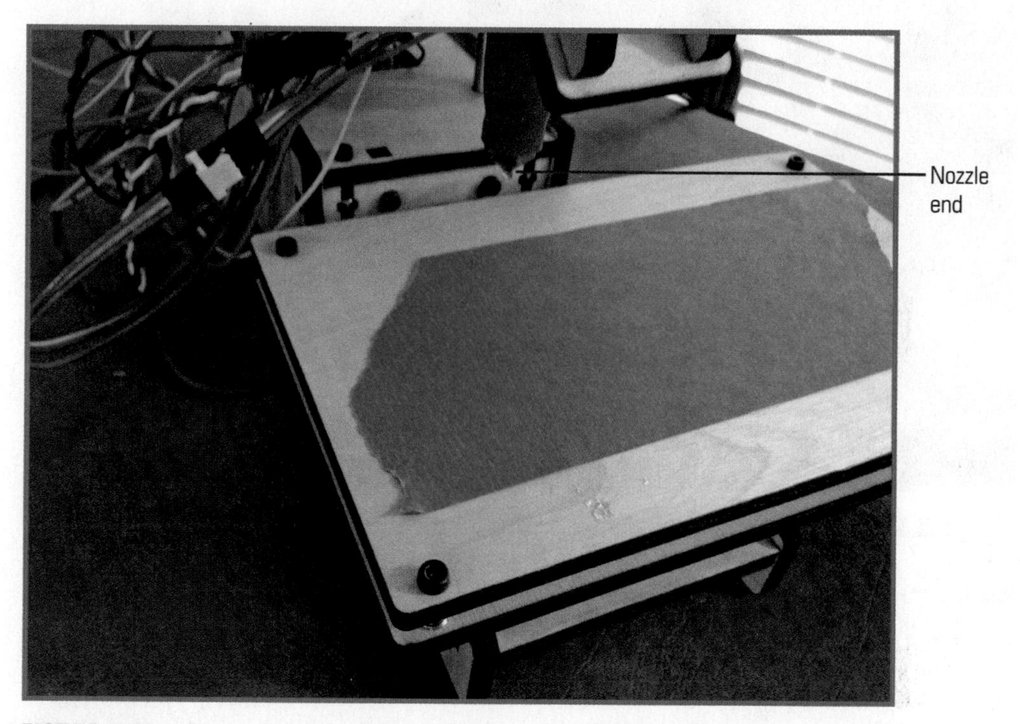

FIGURE 5.10 Treat the tip of the nozzle like a point in space.

When Repetier takes the g-code and begins to issue instructions to the motors, it is assuming that the nozzle tip is in the front-left corner of the print bed and almost touching the print bed. I can manually adjust the X and Y axes on my Simple by carefully pushing or pulling them so that the nozzle is as far forward and to the left as possible (without breaking anything). I can also manually adjust the Z axis by turning the vertical lead screw by hand. If I turn it enough times, I can get the nozzle to the proper starting location, as shown in Figure 5.11.

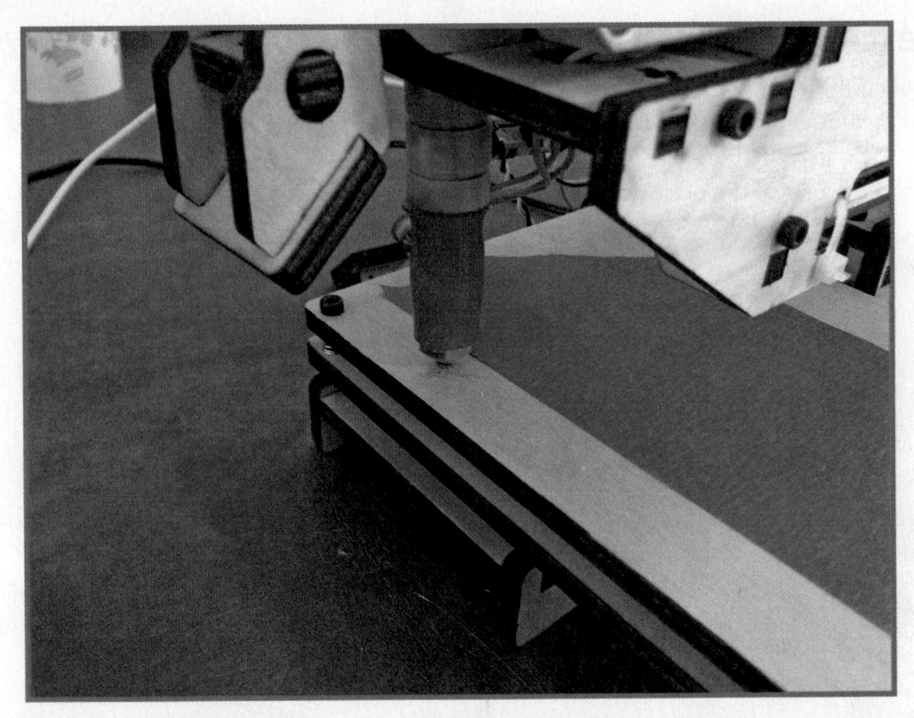

FIGURE 5.11 Nozzle now in proper starting location.

Manually moving the nozzle isn't difficult, but there is a better solution that doesn't require you to adjust the three axes by hand. Change back to the Print Panel shown in Figure 5.12 and note that X, Y, and Z axes each have a collection of strange buttons with values on them: –100, –10, –1, –0.1, 0.1, 1, 10, and 100. These values are in mm, and a single click on one of the axis buttons will cause that motor to move either the print bed or the nozzle that distance.

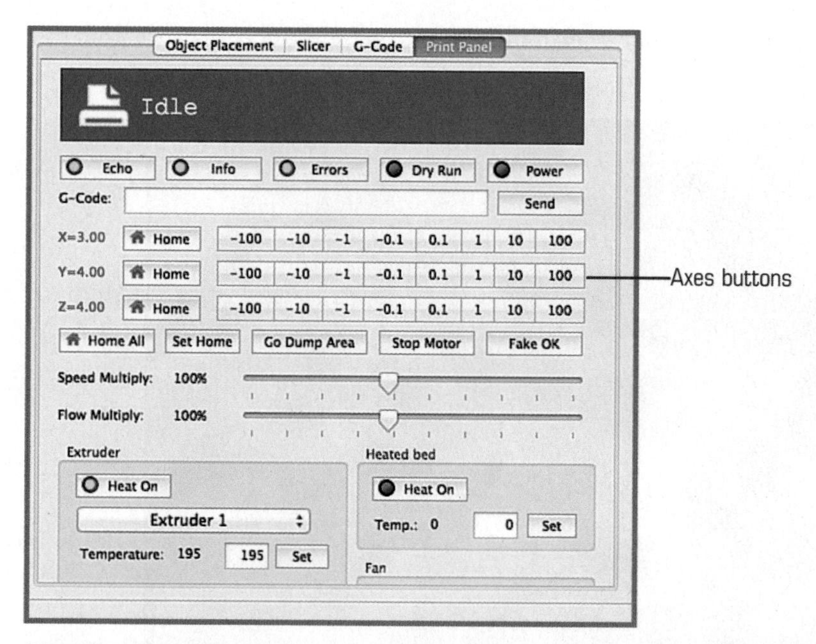

FIGURE 5.12 You can move the hot end nozzle in small or large increments.

To use these controls correctly, you need to determine the direction that a motor will move the print bed or nozzle when each button is clicked. You can do that easily enough using the 0.1 or -0.1 buttons for each axis (or maybe the 1 or -1 buttons). For example, if you click the -1 button for the X axis, the print bed moves 1mm to the right. If you click the -10 button for the X axis, the print bed moves 10mm to the right. If you click the 10 button for the X axis, the print bed moves to the left.

Similar experimenting with the Y axis tells me that the negative buttons (-0.1, -1, -10, -100) move the nozzle forward, toward the front of the print bed. The positive buttons move the nozzle back, away from the front of the print bed. For the Z axis, the negative buttons move the nozzle closer to the print bed, and the positive buttons move the nozzle up and away from the print bed.

As you can see, by properly using these buttons, you can move the print bed and nozzle in such a way that the nozzle's tip is placed in the correct starting position at the front-left corner of the print bed.

One of the secrets to a good print job is getting the nozzle as close to the print bed surface as possible so that it lays down a flat bead of molten plastic during the initial movements. To do this, Printrbot recommends cutting a small square from a sheet of 8.5" x 11" paper. Place this slip of paper under the nozzle tip and lower the nozzle until it requires a slight tug or two to pull the slip of paper out from underneath the nozzle tip. You'll have to experiment a few times until you get the right distance figured out, but as a rule, I lower the nozzle as far as I can get it without it digging into the print bed's surface. Figure 5.13 shows how I've placed this slip of paper underneath the nozzle to get the proper distance from the print bed surface.

FIGURE 5.13 Slip of paper underneath nozzle.

If you find that the slip of paper is too tight and can't be pulled away easily, use the 0.1 button for the Z axis to move it up a tenth of a millimeter, and try again.

When you've got the nozzle in the proper X, Y axis location and the right distance from the print bed surface, tap the Set Home button shown in Figure 5.14.

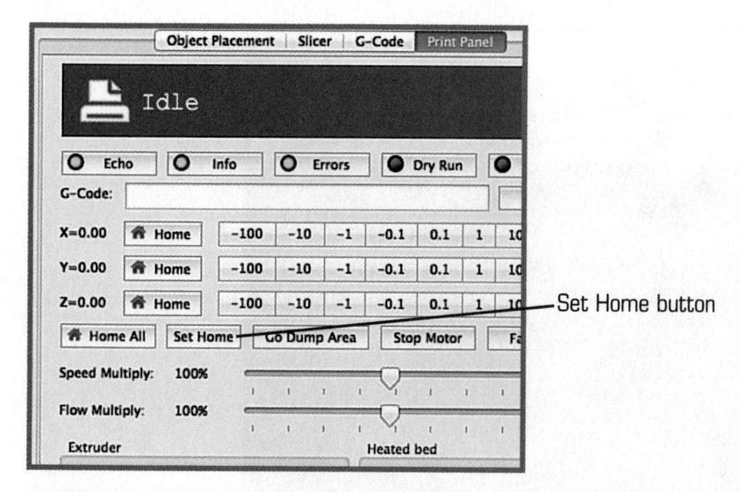

FIGURE 5.14 Press the Set Home button when the hot end is in the right location for starting a print job.

After clicking the Set Home button, you should see that the values for X, Y, and Z on the Print Table change to 0.

Now it's time to print.

Print!

Before clicking the Run button, perform a last-minute check. Switch to the Temperature Curve image and verify that the temperature is where it needs to be. Click the g-code button (to the left of the Print Panel button) and verify there is actual g-code there from the object slicing tasks. Finally, verify that you've homed the nozzle. It should be in the front-left corner and almost touching the print bed surface, and all three axes should have values of 0 on the Print Panel.

Take the end of your filament, clip it off at an angle to provide a sharp point, and feed it down through the extruder and into the hot end. When the filament is inserted properly into the hot end, lock down the extruder to put pressure on the filament so the extruder motor can feed it in on its own.

Now switch back to the 3D View button shown in Figure 5.15 and click the Run button.

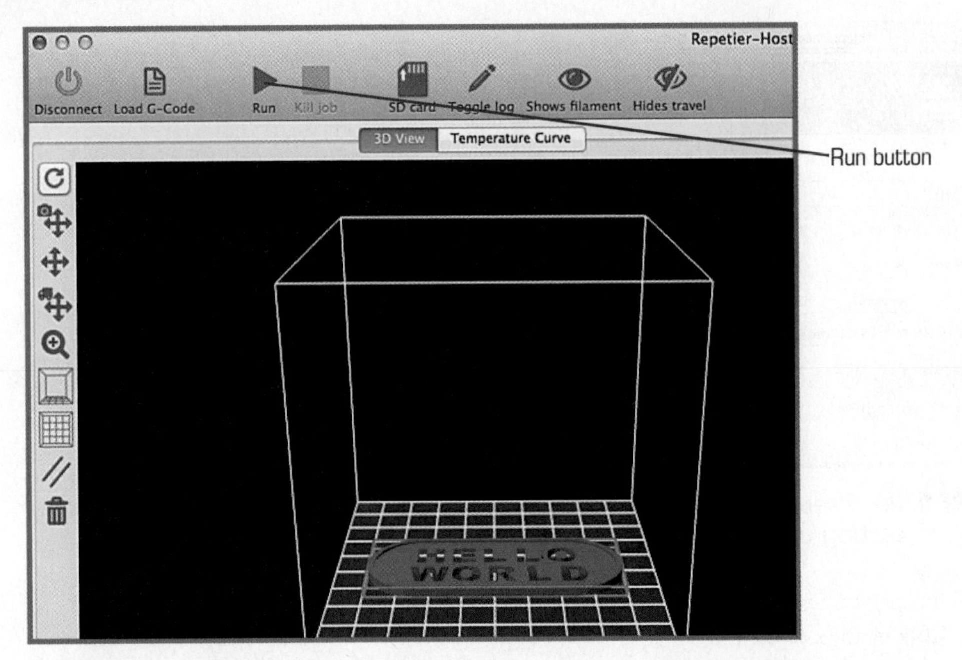

FIGURE 5.15 Watch the progress in 3D view as the path of the hot end is traced.

There will be a slight delay before the nozzle begins to move. When the print job is first executed, the nozzle increases slightly in temperature (up to about 5 degrees or more of the temperature you've set) to get the filament melted and ready to flow.

When the print job begins, the nozzle starts to move. If you've configured Repetier as recommended in Printrbot's "Simple Getting Started Guide," you'll notice that a few loops around your object will first be put down; this is both to get the melted filament flowing through the nozzle and to allow you to visually verify that a bead of plastic is being put down on the print bed, as shown in Figure 5.16.

FIGURE 5.16 An initial bead of plastic is put down around your object.

If you don't see a bead of filament on the print bed, it's possible the extruder is gripping the filament too tightly or too loosely and is not feeding it into the hot end. Loosen or tighten the extruder mechanism on your 3D printer slightly and see if the bead begins. If not, click the Kill Job button and check a couple of things:

- It's possible your hot end might need a slightly higher temperature to get things flowing. Use the Print Panel to raise the nozzle 3cm (30mm) or so away from the print bed. Release the mechanism on the extruder that applies pressure and manually push the filament down into the hot end. You should see a wormy bead of plastic come out of the nozzle, as shown in Figure 5.17. If the bead appears, you've got the temperature set properly and need to adjust the tightness of the feeding mechanism on the extruder.
- If manually pushing the filament down into the hot end doesn't produce a wormy bit of filament coming out of the nozzle, you might need to increase the temperature. Raise the temperature manually by 5 degrees. Manually push in the filament. If it comes out too fast (almost like liquid), pull the temperature back in one-degree increments until you see the filament coming out and immediately cooling and curling up on itself.

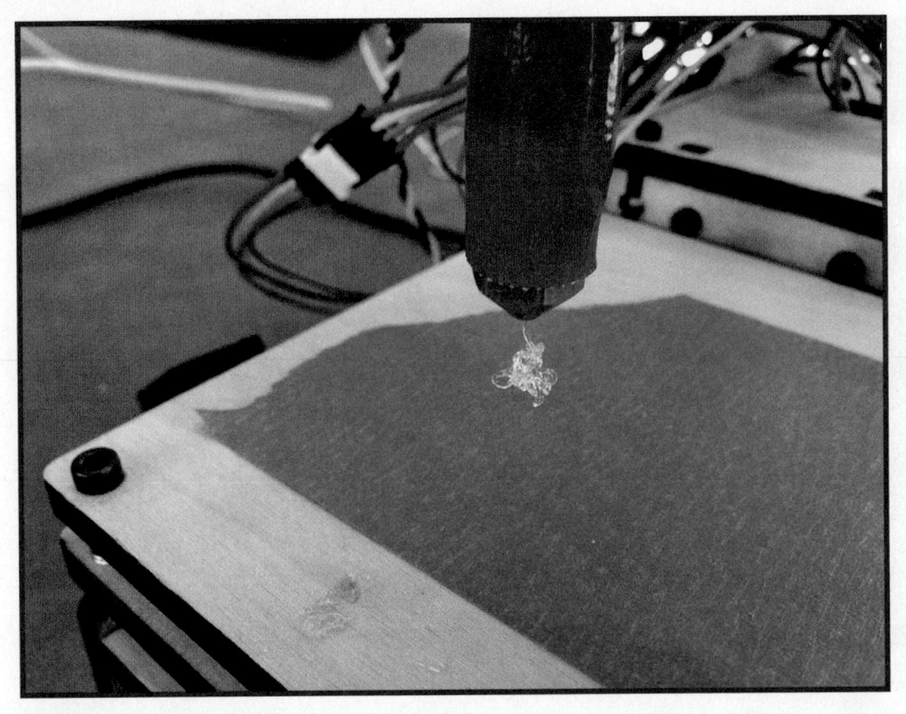

FIGURE 5.17 Testing filament extrusion.

After you've got the plastic extruding properly, you'll have to home the nozzle again and start the print process over (including possibly turning on the heat). You won't have to slice the object again, but click on the G-code button to verify the code is still there.

Your first few (or first few dozen) print jobs aren't going to be perfect, so be aware of that fact to avoid frustration. It took over a dozen adjustments in the nozzle height (above the print bed) and the tightness of the extruder feeding mechanism to get it "just right" before the Hello World medallion began to take shape, as shown in Figure 5.18.

FIGURE 5.18 The Hello World medallion printing properly.

Notice in the photos that I've put down a wide strip of blue painter's tape on the print bed as I wrote about at the beginning of the chapter. I did this before the print job began (and before I turned on the heat). Don't forget the painter's tape—it will help you remove the finished print job—you don't want the hot plastic stuck to the wooden print bed! Just peel the tape off the print bed when the print job is done, and your printed object should pop off the tape easily. You can see the finished medallion (before removing from tape) in Figure 5.19.

FIGURE 5.19 The finished medallion.

Upgrades!

Homing the nozzle for every print job can get annoying fast, but there's an upgrade that now ships with every Printrbot Simple. It's a set of end stops that automatically do the homing for you—cool! The end stops were released too close to the finishing of this book, so I was unable to include their installation in Chapter 3, "Assembly Assistance for the Printrbot Simple." But Printrbot has added the steps to install the end stops, and I highly encourage you to use them. After installation, instead of using the axes buttons to move the nozzle tip to the proper starting location and distance from the print bed, you click the Home All button on the Print Panel and this task is done automatically. The motors spin and the print bed and nozzle move on their own, and no fine-tuning by you is needed.

As for the painter's tape, there's also an upgrade coming soon (but not ready at the time of this writing) that will eliminate the need for the tape. It's called a heated bed, and it's a tool used to let the plastic cool gradually so that it doesn't warp or curve. If you're a painter's tape user, you may notice that the plastic sometimes has a tendency to warp slightly as the plastic cools. The heated bed helps prevent his. Be sure to check with your 3D printer manufacturer to see if a heated bed is available; if you're the owner of a Simple, the heated bed should be available by the time you're reading this.

Free 3D Modeling Software

Chapter 5, "First Print with the Simple," demonstrated how easy it is to print a 3D object with the Simple. This process is fairly typical of most 3D printers and other 3D printing software you may use. I briefly introduced you to Thingiverse, a library of 3D objects that you can search and download the files needed to print those objects. Many 3D printer owners are as happy to just download free objects from Thingiverse and other repositories, but what if you're interested in creating and printing your own custom 3D models? For creating your own 3D models, you're going to need the special CAD software I mentioned earlier in the book. There are a lot of CAD software companies and even more CAD applications available. Some are extremely advanced CAD applications with hundreds of features and capabilities, and some are super-simple ones that have some limitations. Some CAD software companies offer training to show users how to properly use the software, and some are written by one individual, after hours, and come with almost no documentation at all. Prices can range from the super expensive to the completely free.

Fortunately, beginning 3D modelers do not need to spend a lot of money on CAD software. Some of the best stuff out there is not only free but also comes with a decent amount of tutorials or training material. However, designing objects with CAD is a skill, and your success (or failure) with a particular CAD application is going to depend on how much time you put into exploring, experimenting, trying new features, and so on. You may find as you develop your CAD skills that a particular application isn't growing with you. When that happens, it may be time to investigate more powerful CAD applications.

In this chapter, I introduce you to a free-to-use "beginner" level CAD application called Tinkercad. (I put beginner in quotes because some CAD modelers can do amazing things with this application.)

Tinkercad is only one example, and you may already have a favorite or a CAD app that a friend has introduced to you that you enjoy using. That's great! Just keep in mind that with any CAD application, you'll want to make certain the application has the capability to save or export an STL file. STL files are the secret ingredient for printing 3D objects with a 3D printer, so if that's your final goal, find and use a CAD application that can take your finished 3D model and save it as an STL file.

Tinkercad

Tinkercad is owned by Autodesk, a leading CAD application developer. The good news is that it's free to use, although there are pricing plans that provide additional features.

Tinkercad is a web-based application, which means you need a web browser. But not just any web browser—it must support WebGL, a special bit of software that gives a web browser the capability to do special things with 3D objects onscreen (among other things). Not all web browsers support WebGL, so you can find out quickly by opening your favorite web browser and pointing it to tinkercad.com. If you see a message like the one in Figure 6.1, you'll need to download and install a different browser if you want to use Tinkercad.

FIGURE 6.1 Error message indicating WebGL is not supported.

Four free web browsers that do support WebGL are Opera (www.opera.com), Internet Explorer 10 (http://microsoft.com), Firefox (http://www.getfirefox.net), and Chrome (https://www.google.com/intl/en/chrome/browser/). There is nothing wrong with installing two or more web browsers on a computer, so feel free to continue using your favorite web browser if it doesn't support Tinkercad. Just open up Firefox or Chrome when you want to use Tinkercad.

I show you Tinkercad and point out some of its features in the remainder of this chapter. I really enjoy Tinkercad because I, too, am a novice 3D modeler and find Tinkercad to be very friendly and non-overwhelming when it comes to buttons and features. And I am slowly but surely learning my way around another free AutoDesk application called 123D Create (123dapp.com) and anticipate moving over to it completely as soon as my skills have increased and I find myself able to create more advanced models.

So, let's take a look at Tinkercad. Its name is a play on the famous toy for children, Tinker Toys, which lets kids build more complex objects by attaching simple shapes (like circles and round beams) together. You'll find that Tinkercad takes that same concept and applies it to creating 3D objects.

Examining Tinkercad

For my discussion on Tinkercad, I've taken screenshots from the Chrome web browser. As you can see in Figure 6.2, when you first visit tinkercad.com, you're given a chance to create a user account. It's free to create an account and absolutely required to use the tool, so go ahead and create one if you want to follow along.

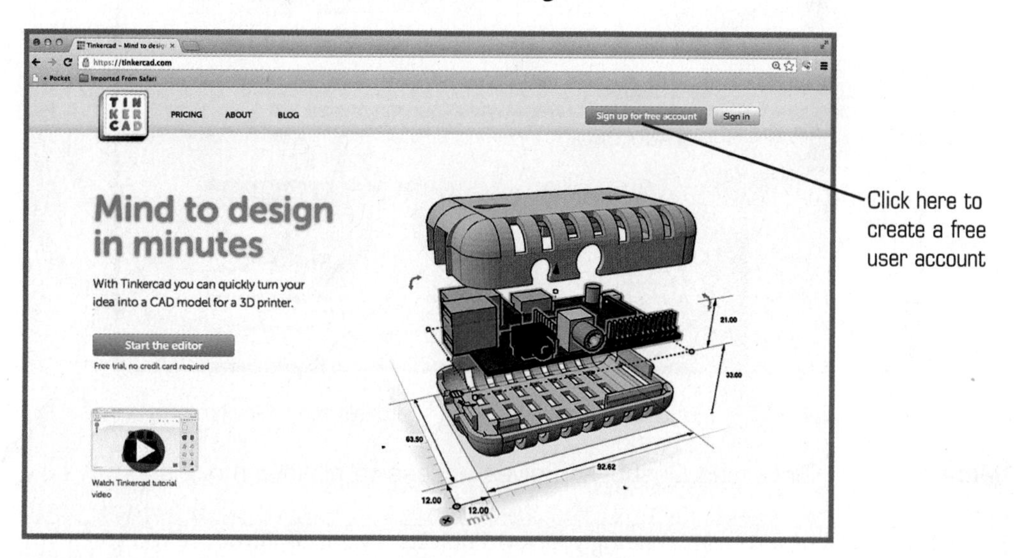

FIGURE 6.2 Create a free account to start using Tinkercad.

In Figure 6.2, notice the unusual 3D object displayed on the right. I'm hoping that single image gives you an idea of the power of Tinkercad. Someone has modeled (drawn) a circuit board that will be protected by a shell consisting of an upper and a lower piece. It's those two pieces, upper and lower, that could easily be printed on a 3D printer!

NOTE

Many books and online tutorials can train you to create these kinds of models. Don't be intimidated when you see advanced 3D models. As you'll learn as you invest more time into learning a CAD application, the objects you can design yourself will advance in complexity as your own skills increase. I include an appendix at the end of the book that lists various CAD applications, as well as books and online resources.

After you've created an account, go ahead and log in. After you log in, you'll see a screen similar to the one in Figure 6.3. This is your personal Dashboard where you can view the models you've created (or are working on), the tutorials (they're called Quests) you've finished, and objects that you've modified from other users (more on that in a moment).

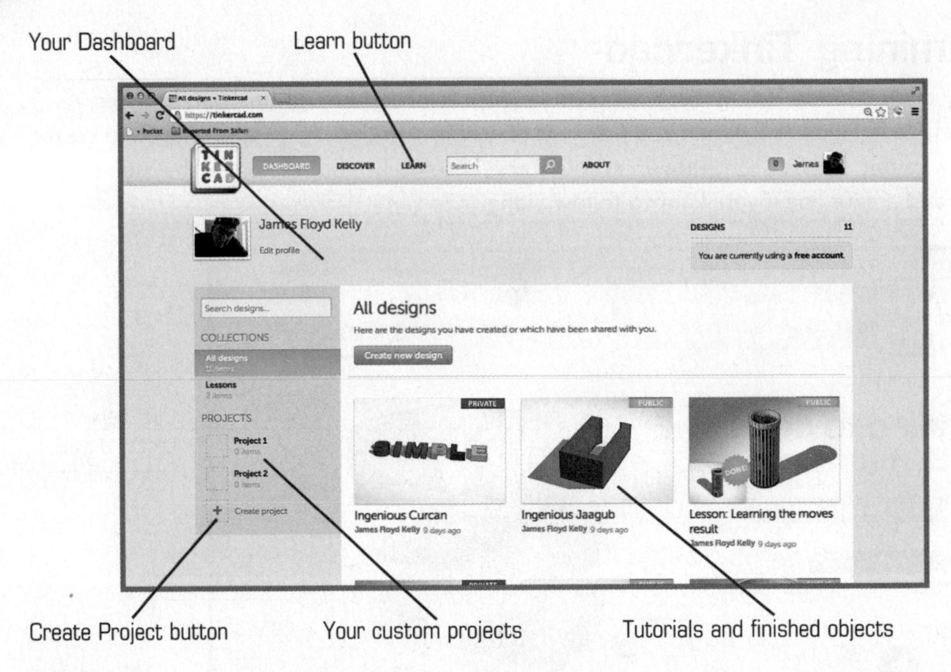

FIGURE 6.3 The Tinkercad Dashboard gives access to all your models.

Before I show you how to start creating your own objects, I want to introduce you to Quests. These are simple tutorials that explain some of the user interface elements. I highly recommend that you work through them all. To view them, click the Learn button at the top of the screen. Figure 6.4 shows the new screen that appears, listing all the Quests.

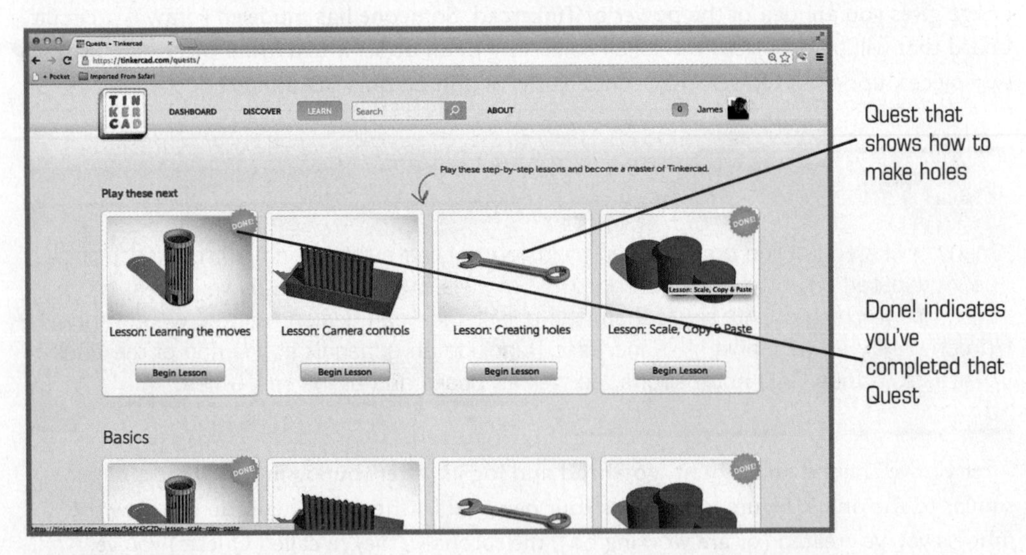

FIGURE 6.4 Complete Quests to learn important skills in Tinkercad.

As you walk through a Quest, you'll be given instructions in a small gray square with a Next button that you click to move to the next step. Go through as many as you can; the lesson called Die on a Workplane is one I found to be very helpful when using Tinkercad because it shows you how to focus on a specific surface of a 3D object, but all the Quests are valuable and should be attempted.

After you finish the Quests, you're likely to want to stretch your wings and start using a blank work surface to create something of your own. I do just that in Chapter 7, "Creating a 3D Model with Tinkercad," but for the remainder of this chapter I want to point out some of the tools that you'll find the most helpful as you use Tinkercad.

First, on the Dashboard, click the Create Project button (refer back to Figure 6.3), and you'll be given a chance to write up a description of your project (click the Add a Description link) before clicking the Create New Design button. See Figure 6.5.

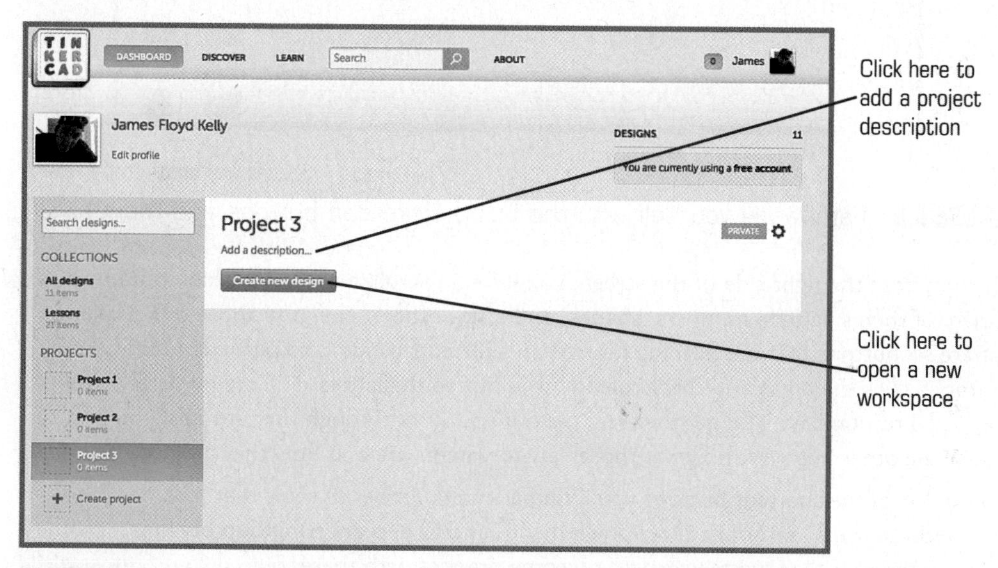

FIGURE 6.5 Name your project, and writeup a small description.

After clicking the Create New Design button, you'll be provided with a completely blank workspace, as shown in Figure 6.6. Running down the right side of the screen is a toolbar; running horizontally across the top of your workspace are a few menu items and tools that I explain next. You may have noticed that Tinkercad provides a funny and unusual name for your project, but don't worry—I show you how to change that later in the chapter.

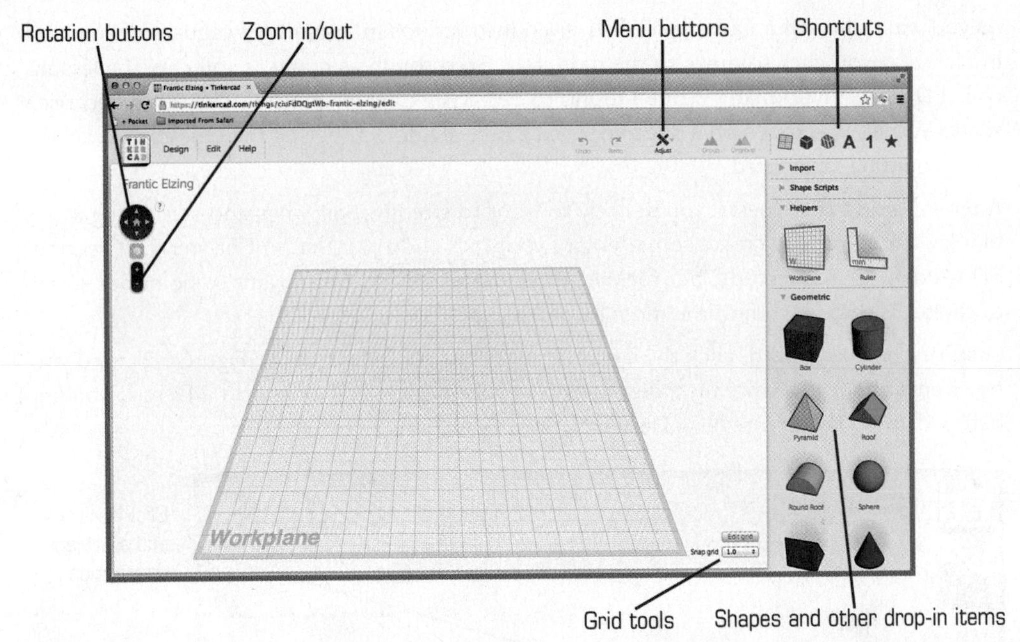

FIGURE 6.6 Familiarize yourself with the basic Tinkercad buttons and menus.

Starting from the right side of the screen, you'll find a scrollable window that contains a variety of things: letters, numbers, shapes, and a few other items. Just above this scrolling list are six buttons that are nothing more than shortcuts to jump to particular items. For example, click the big A and the scrolling list jumps to the letters. The big numeral 1 will jump you to numbers. You get the idea. (And if you work through the tutorials, you'll learn about the other buttons and what those category items are and how they are used.)

To the left of the shortcut buttons you'll find a small number of tools that include Undo and Redo buttons, an Adjust drop-down menu, and Group and Ungroup buttons. Again, working through the Quests will give you some practice with these, but I also use most of them in Chapter 7 and explain them in more detail there.

At the bottom of the screen below the Group and Ungroup menu buttons is the Edit Grid button and the Snap Grid setting. These two items are very useful for the accuracy of designed objects, and I show you in Chapter 7 how to use the Edit Grid button to change the size of the workspace so it matches your 3D printer's print bed size; this way, you'll know that whatever you're designing will fit within the allowable X and Y print area (length and width) of your 3D printer. You will need to monitor the height of any objects you design to make sure they are no higher than the maximum Z axis print size of your 3D printer. Right now, there is no way to limit the height of an object created in Tinkercad.

In the upper-left corner of the screen are a few more buttons that include Design, Edit, and Help. The Edit drop-down menu offers Copy, Paste, Duplicate, and Delete options (but keyboard shortcuts make performing these four tasks much faster). The Help menu provides access to a basic video tutorial and a Learn More About Tinkercad option. Unfortunately, neither of these are all that helpful; hopefully, AutoDesk might add some Help files and additional videos over time—and you're likely to find more help by doing a Google search for Tinkercad assistance.

Creating simple objects in Tinkercad doesn't get much easier. You drag and drop basic shapes on the workspace, as shown in Figure 6.7. Simply click and hold on a shape from the scrollable window on the right (such as the Box) and drag it to the workspace before releasing your mouse button.

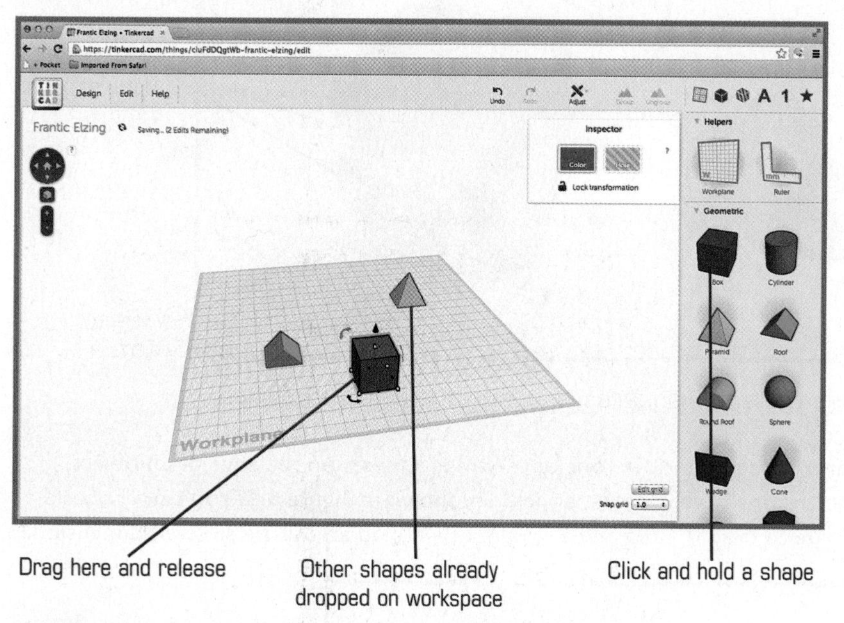

FIGURE 6.7 Drag and drop objects on the workspace.

Shape sizes and orientations are modified by clicking an object once to select it. As you can see in Figure 6.8, an object displays certain smaller icons that are used to resize, rotate, and perform other tasks. (Note that I've zoomed in on the cube for greater detail—use the + and - buttons on the left side of the screen to zoom in and out.) To resize an object, you click and hold the small white squares that appear on two edges and on the top surface of an object and move them to adjust a dimension. The small white square on top of an object lets you increase or decrease the height of an object.

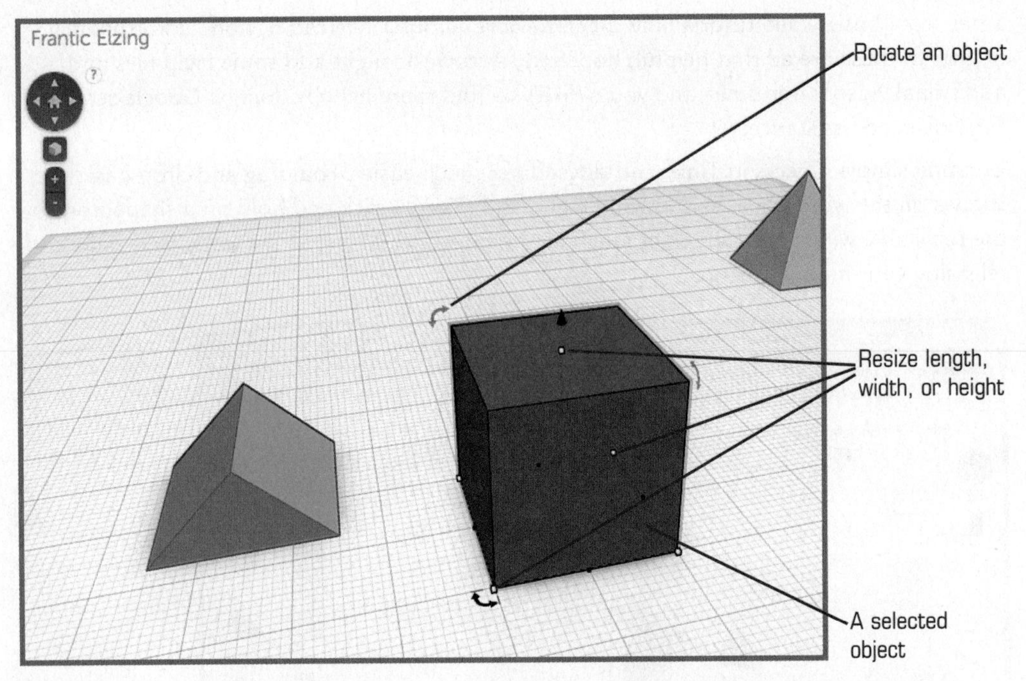

FIGURE 6.8 Use the small squares to resize the selected object.

If you place your mouse pointer over one of the small squares on the edge of an object, you'll see the dimensions of that object appear, as shown in Figure 6.9. You can click and hold a small square to drag it; dragging a square will let you adjust the selected dimension.

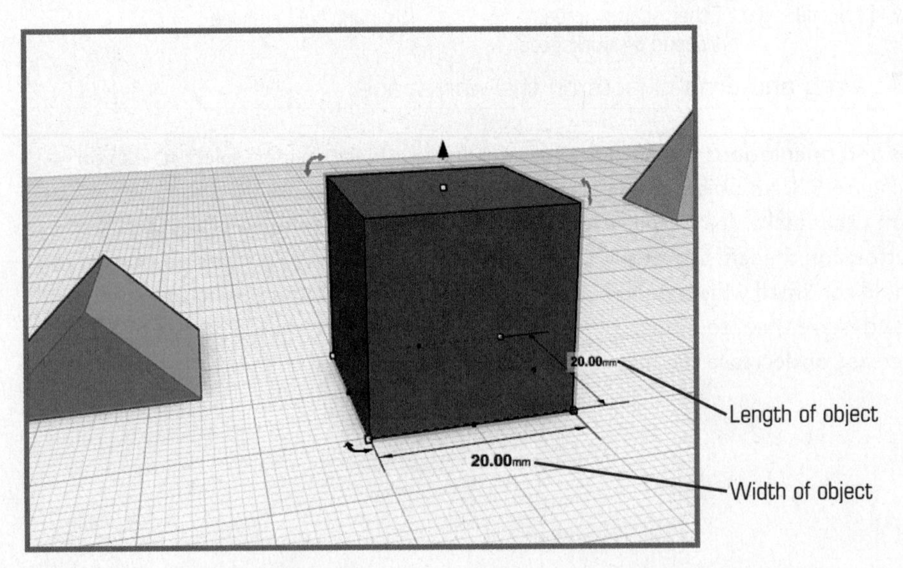

FIGURE 6.9 Change the width and length of an object.

If I hover my mouse pointer over the small white square on top of an object, I can view the object's height. If I want to increase the height of that object, I click and hold that small square and drag it up. Dragging up increases the height, and dragging down decreases the height. Figure 6.10 shows that I've increased the height of the box.

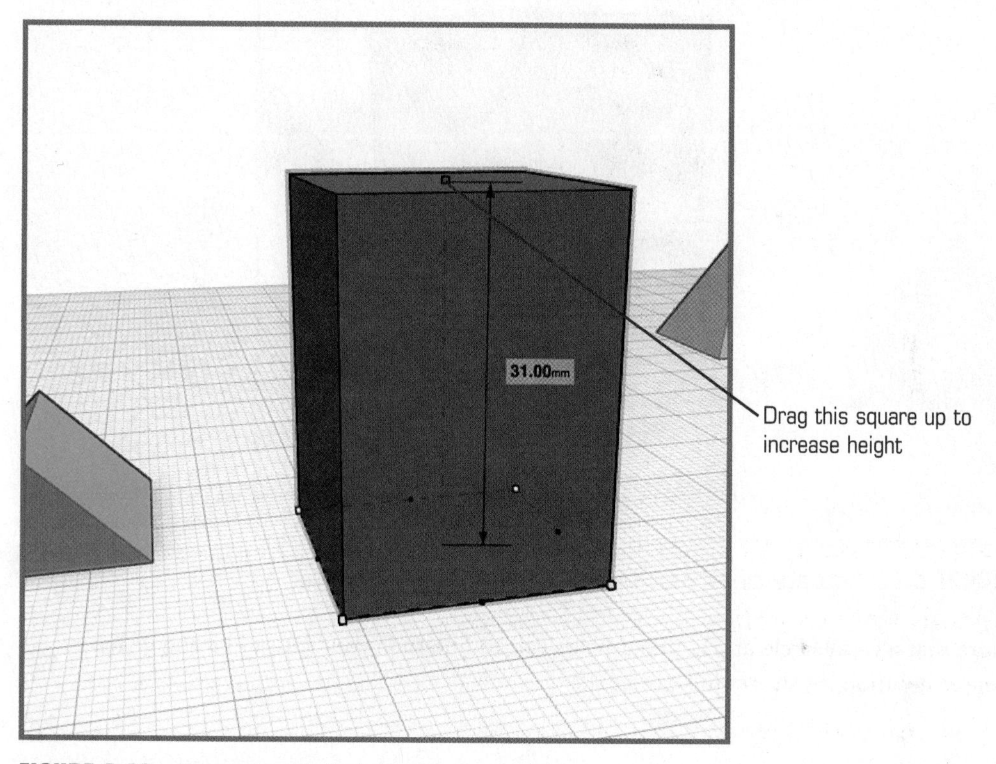

FIGURE 6.10 Change the height of an object.

Rotating the object left or right involves clicking and holding on the small circular arrow shape indicated in Figure 6.11. Using this arrow, you can rotate an object on the Z axis. You'll see two other similar circular arrows near the top of the object—each lets you rotate the object on the X and Y axes.

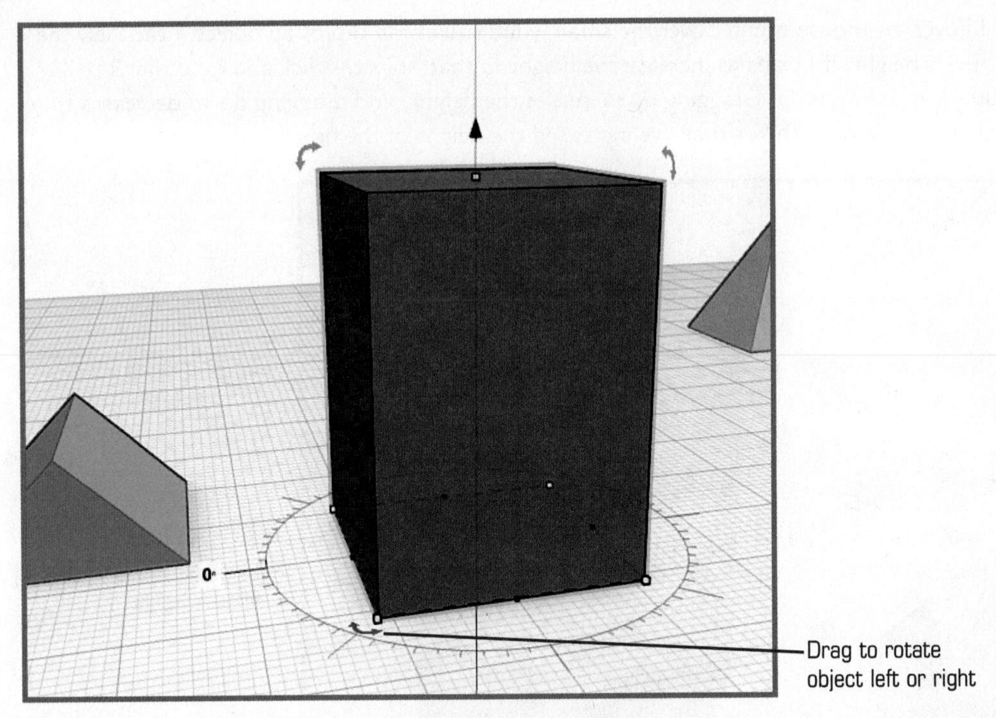

FIGURE 6.11 Rotate an object by using the circular arrows.

Note that a small circle appears that helps you to fine-tune your rotation using 0–360 degree notation, as shown in Figure 6.12.

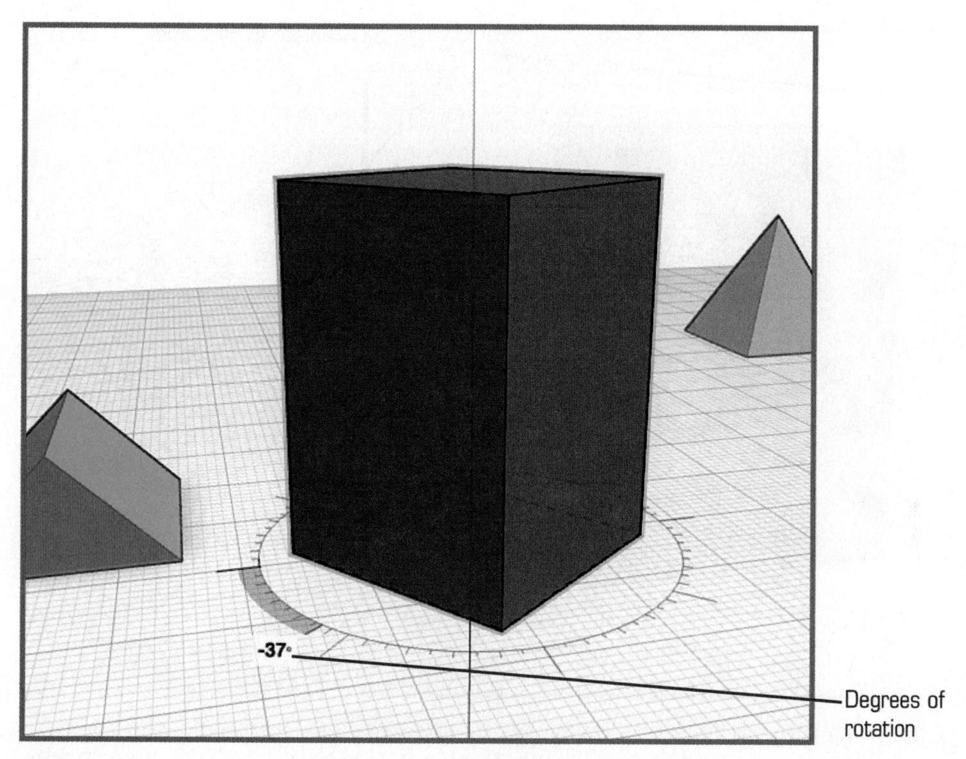

FIGURE 6.12 Fine tune a rotation in 1-degree increments.

Two other circular arrow controls are available, as shown in Figure 6.13. Both allow for additional rotation (if you want to flip an object upside down, for example). Depending on which you choose, the object rotates with respect to the X axis or Y axis. As with rotating an object left or right, you'll be given a numeric indicator that will help you fine-tune the rotation angle.

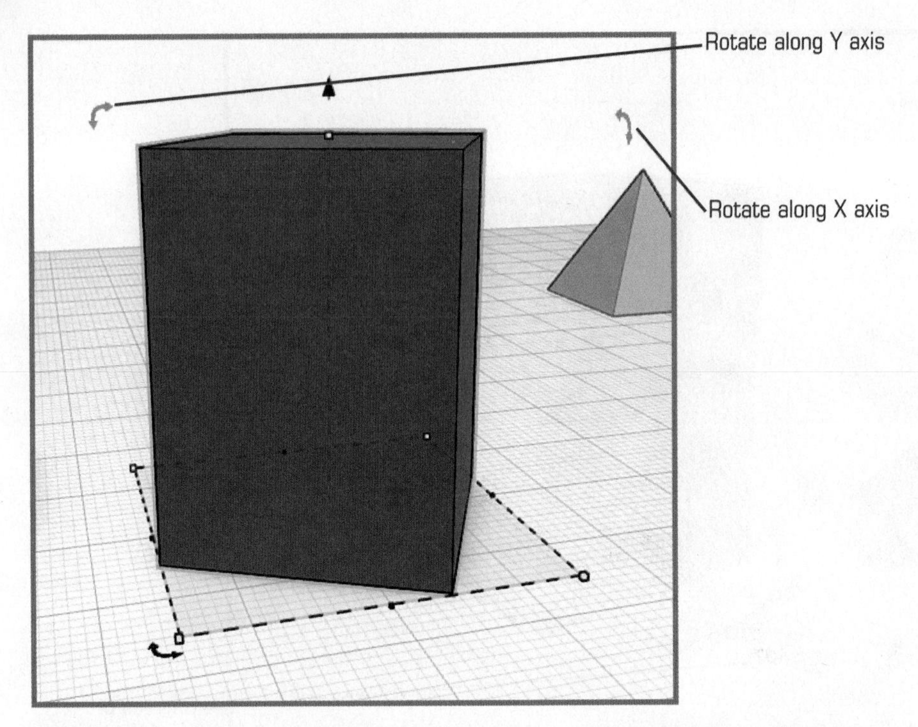

FIGURE 6.13 Rotate an object in 1-degree increments along X or Y axes.

If you want to increase or decrease the size of an object but keep its dimensions at the same ratio, you can hold down the Shift key before you click one of the small squares. Then, when you drag your mouse pointer to adjust the size, all dimensions (height, width, and length) are "locked" and adjust at the same ratio. Figure 6.14 shows that I shrank a Box object (that's also a Cube because its three sides are of equal value) by holding the Shift key down while dragging the width box. As you can see, the cube is tiny, but the height, width, and length all stayed equal in value. If I hadn't held the Shift key down, the width would have shrunk but the height and length wouldn't have matched.

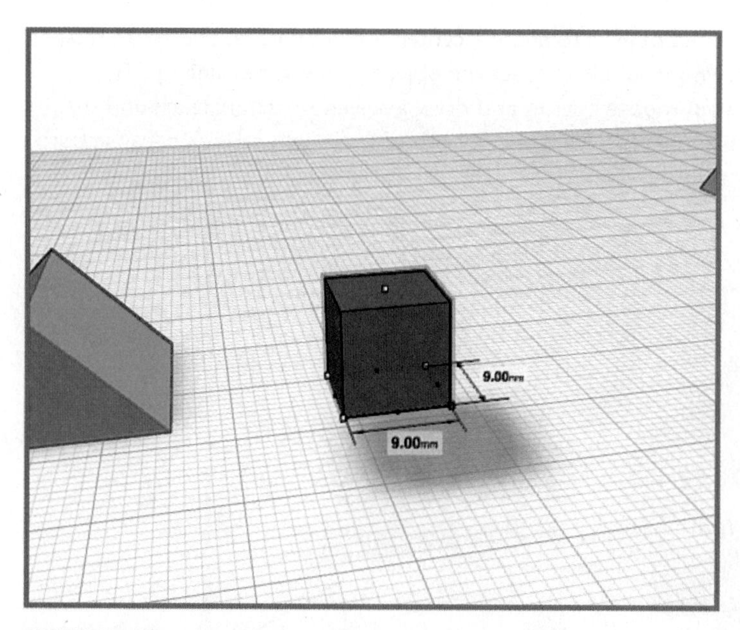

FIGURE 6.14 Use Shift to shrink or enlarge an entire object at a consistent rate.

Take a look at Figure 6.15 and you'll see two triangular objects. Notice that they are not centered together (where their center lines match up) nor do they share a similar edge (where one edge on each object lines up).

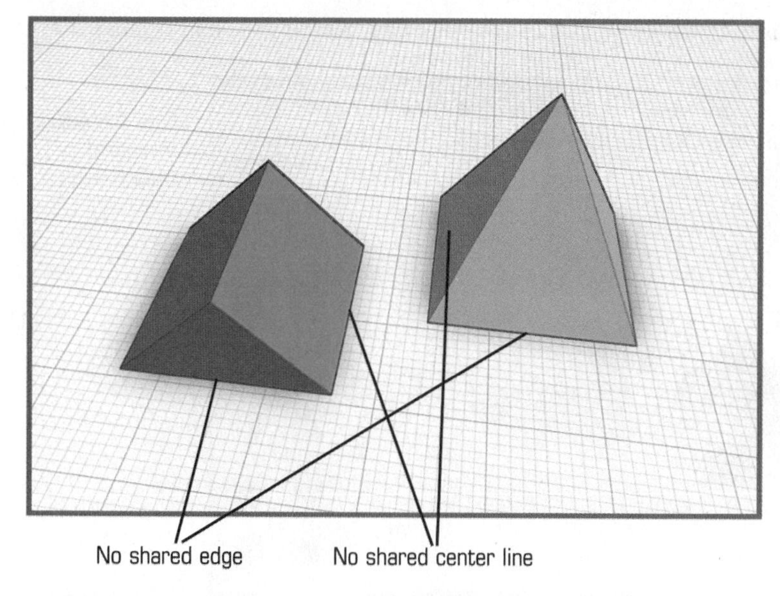

FIGURE 6.15 There's an easy way to line up objects.

You'll have times when you need objects to share a center line or an edge, and to do this, you use the Adjust drop-down menu. First, select the objects you want to line up. To do this, press and hold down your mouse button and draw a selection rectangle around the objects. As you can see in Figure 6.16, I've selected both objects—you can tell because both objects are surrounded by a dotted line.

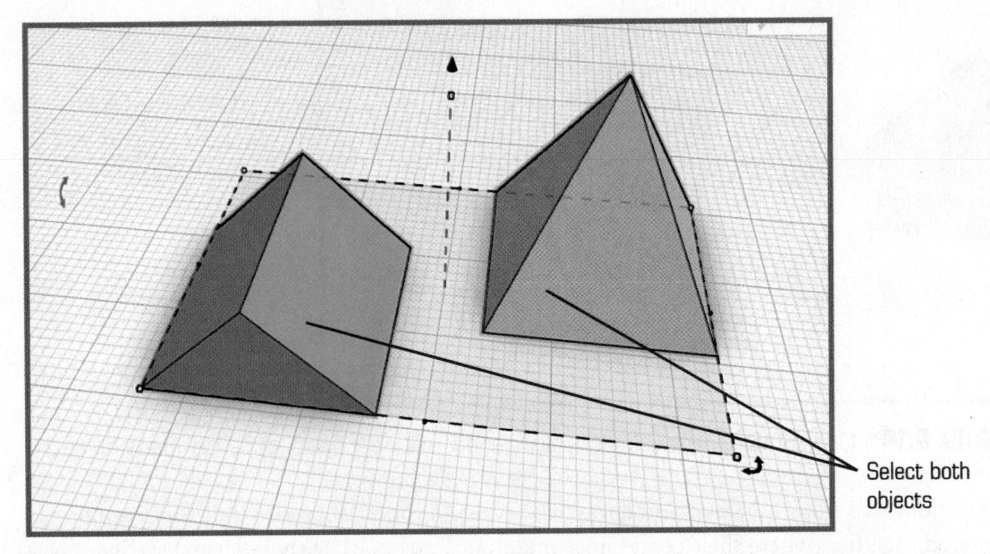

FIGURE 6.16 Select all objects that you wish to share an edge or center line.

Next, click the Adjust drop-down menu and choose the Align option shown in Figure 6.17.

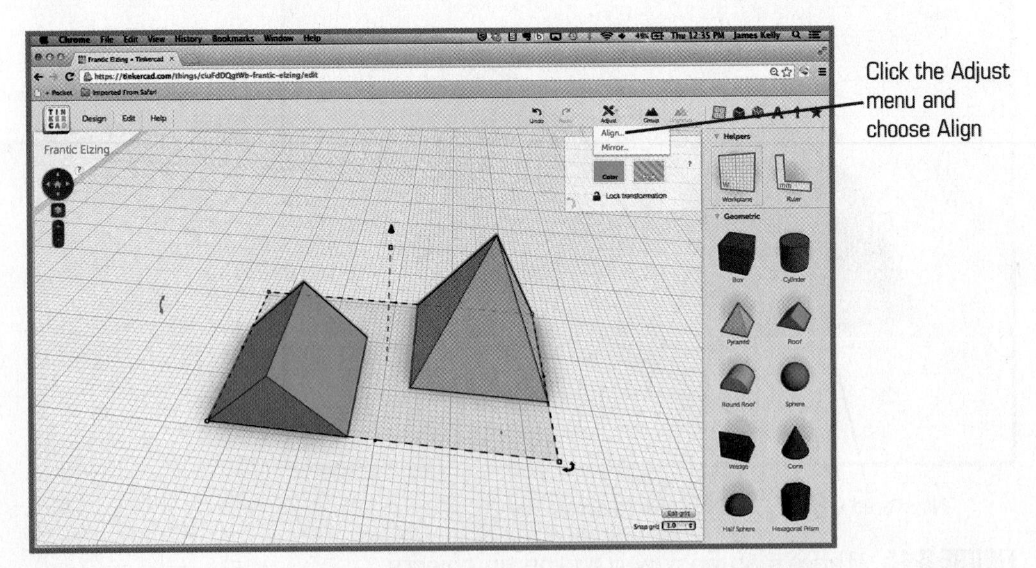

FIGURE 6.17 Use the Adjust menu to choose an alignment for selected objects.

The Align option puts a number of lines and dots on the workspace, as shown in Figure 6.18. (These can be a little tricky, so I encourage you to drop in a couple of shapes and try out this Align feature to see how it works.)

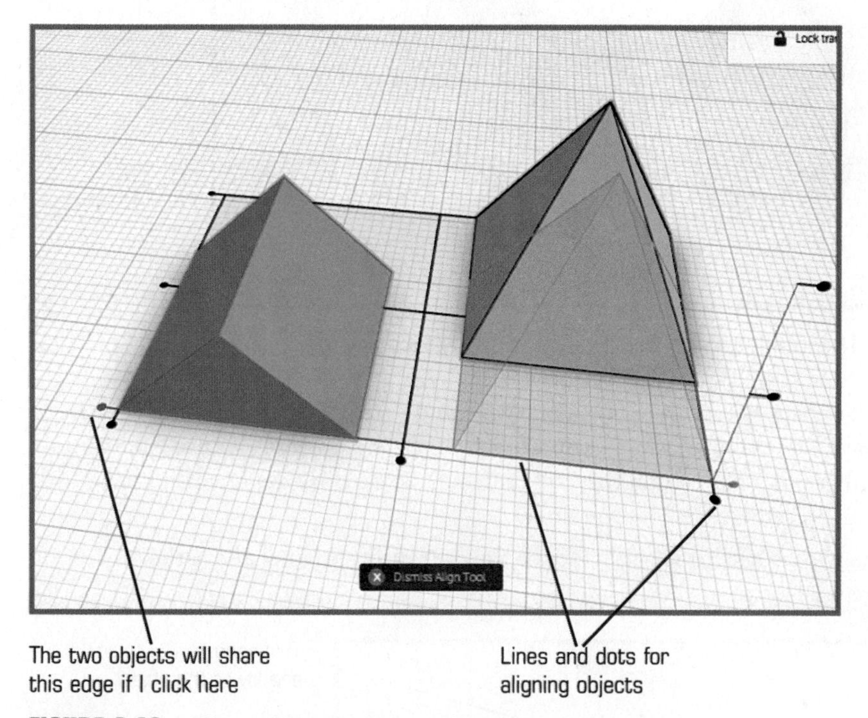

FIGURE 6.18 Align objects along edges or center lines.

These lines and dots represent different edges that can be used to align the selected objects. For example, if you want to have the two objects lined up using their front edges, you can hover your mouse pointer over the dot indicated in Figure 6.18. When you do this, you see an outline of the triangular object on the right that has been shifted to align with the front edge of the object on the left. If you choose to implement the suggested position change, click the dot and the change is made, as shown in Figure 6.19. (Note that you can click the Undo button once and the change is reversed.)

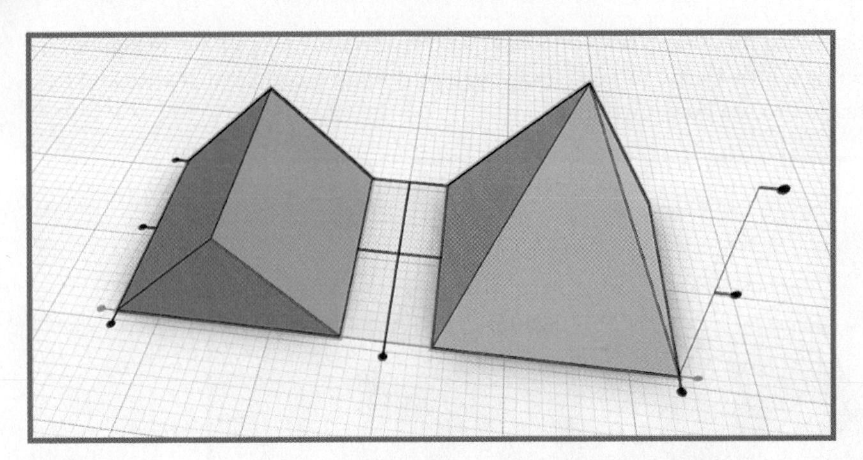

FIGURE 6.19 The objects are lined up along a shared edge.

The Align tool is very useful. With it, you can have two different shaped objects line up in a number of ways. For example, what if you want to stack two different shaped objects on top of one another but make certain they are centered? To do this, you first need to adjust the height of one object. Here's how you do it.

Click one of the two objects to select it. Notice the dark arrow floating above the object shown in Figure 6.20.

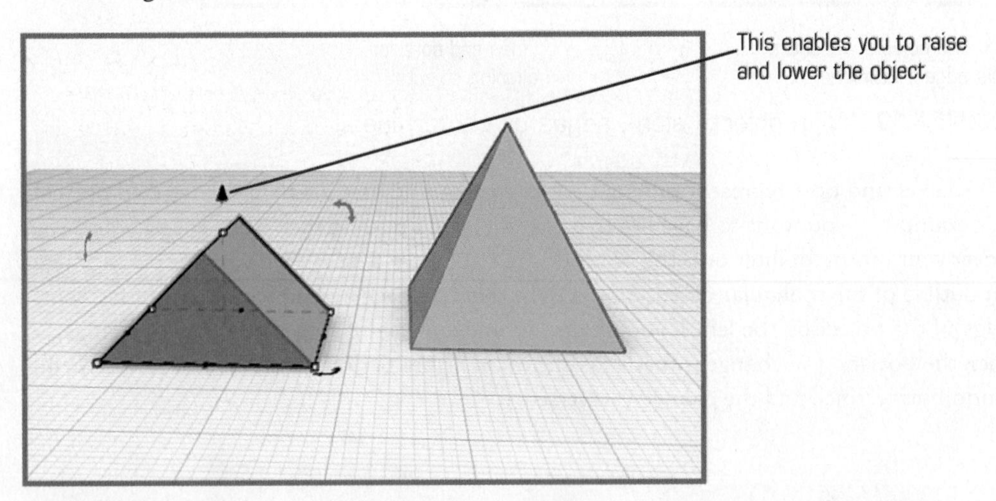

FIGURE 6.20 Raise and lower an object when you wish to stack objects.

That arrow enables you to adjust the position of the object above or below the workspace. Try it out! Drag it up and the selected object floats up above the workspace. Notice in Figure 6.21 that as you drag, you'll see a changing value that indicates how far above or below the object is relative to the flat work surface. (You'll also notice a shadow effect below the object that lets you know it's floating up higher.)

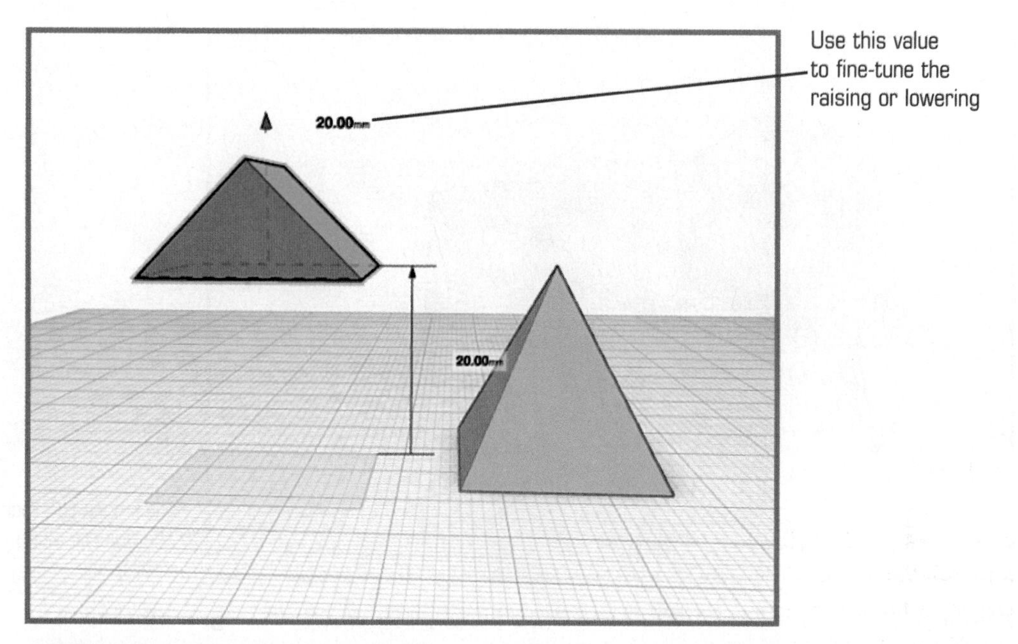

FIGURE 6.21 Use the height value to determine the proper raising and lowering of an object.

With one object above the other, you can use the Align feature to stack them perfectly. First, select both objects so you can view the control dots (refer back to Figure 6.18). Notice in Figure 6.22 that the workspace is rotated so you are looking slightly down on the two objects. Hovering the mouse over the specified dot will show you what the stacked objects will look like. If you like it, click the dot to make the change.

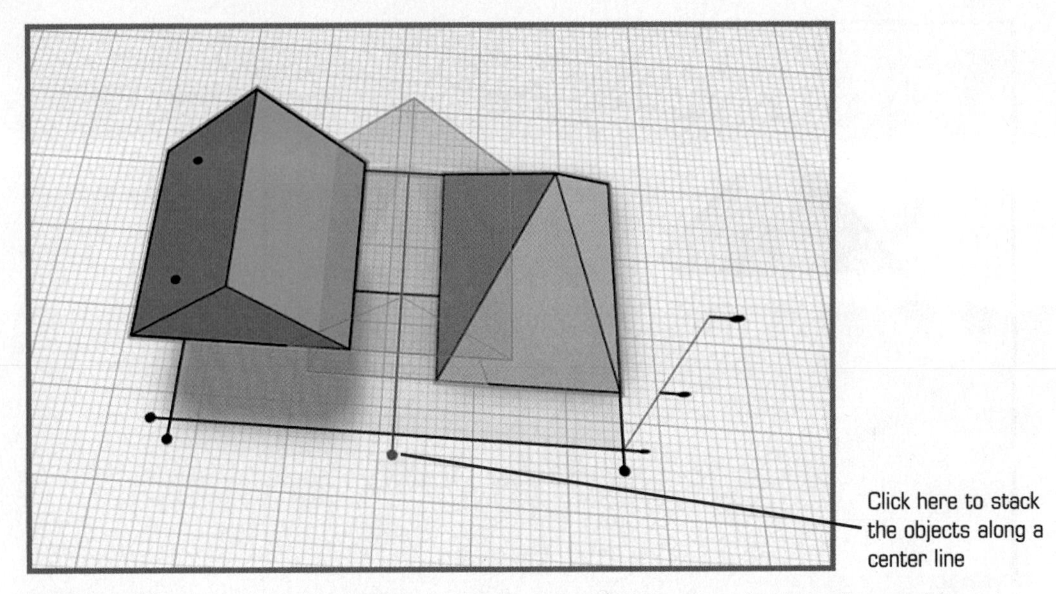

FIGURE 6.22 It is easier to confirm centered objects when looking from above.

The two objects are centered along one center line. If you rotate so you are looking from slightly above and to the front of the top object, as seen in Figure 6.23, you will see that the objects are centered down the selected center line.

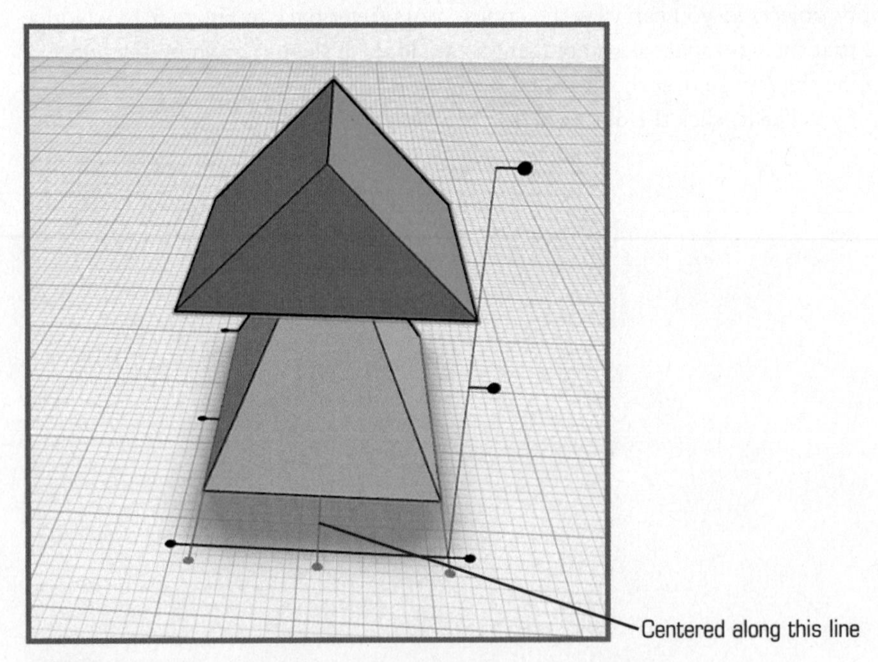

FIGURE 6.23 The objects are centered along one axis.

Although the objects now appear centered, it's not centered along the other axis. You can rotate the view so you are looking at the objects from the side to verify this. Figure 6.24 shows the result.

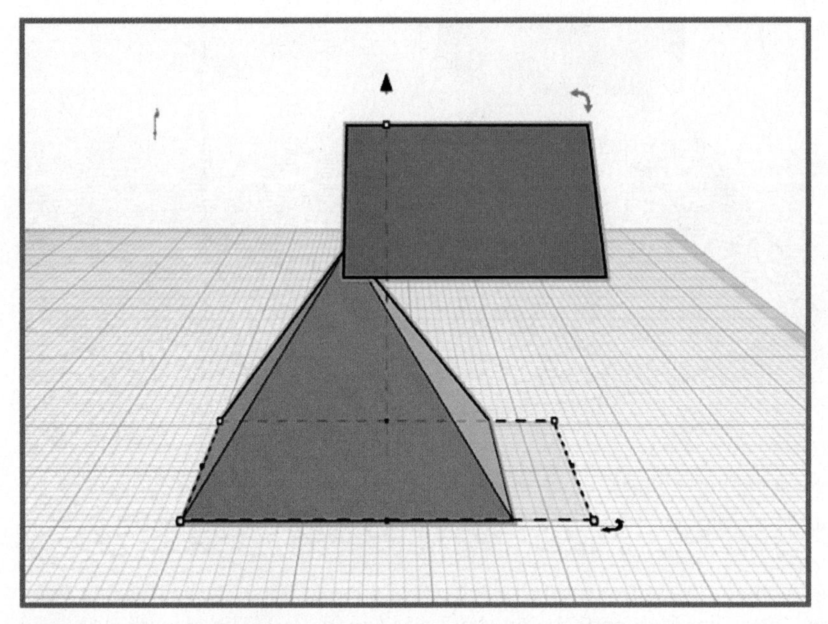

FIGURE 6.24 Not centered along the other axis.

As you can see, they may be centered when looking from the front, but they are not centered from the side. But that's an easy fix. Select both objects, click the Adjust menu, and select Align once more. Click the dot indicated in Figure 6.25 to center the two stacked objects along this center line.

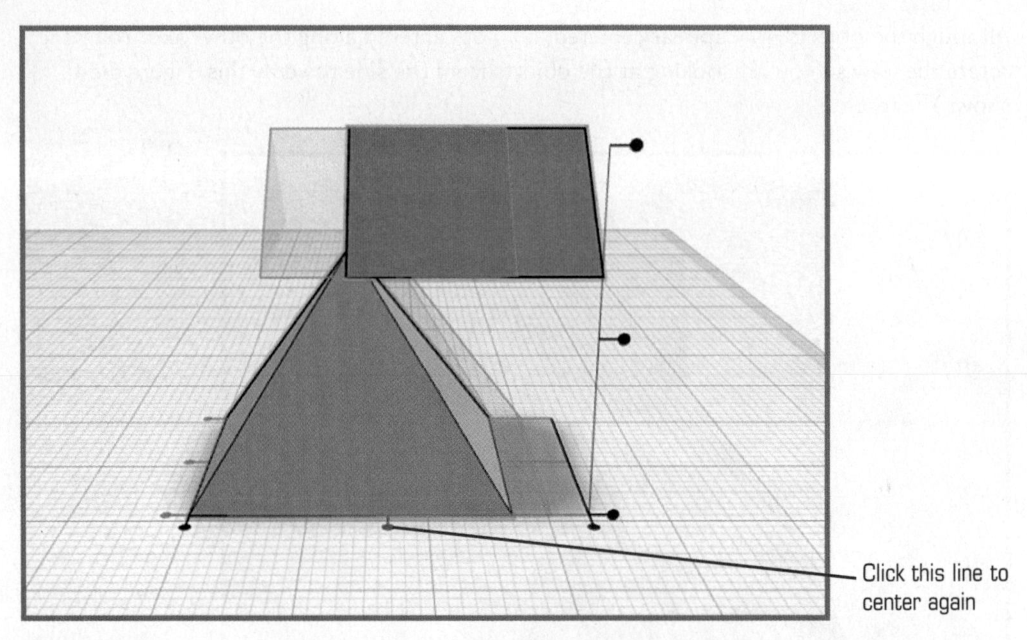

FIGURE 6.25 One final alignment is needed to stack the objects perfectly.

The result is shown in Figure 6.26. I've rotated the camera view so you can see that the two stacked objects are indeed lined up perfectly on the Z axis (up/down). The camera view is looking up from underneath the flat workspace; imagine the workspace as a glass table surface that you can see through, and you'll get the idea.

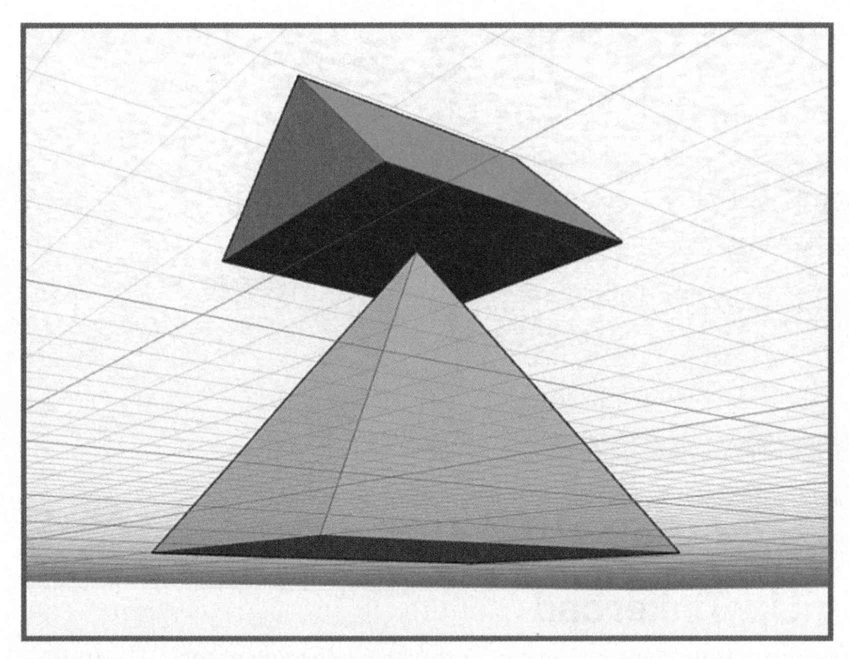

FIGURE 6.26 Perfectly centered.

You can use a number of other tricks (such as the Align method for centering objects) to create 3D models within Tinkercad, and I show you more in Chapter 7 when I create an object to print out with the Printrbot Simple.

Keep in mind that when you create an object with Tinkercad, your object is saved if you close down the web browser; it'll be there waiting for you when you next open the app and log in to your Dashboard. But as good practice, always leave Tinkercad by clicking the Design drop-down menu and selecting Save. You can also use the Properties option to change the name of your file, as shown in Figure 6.27. You can select Public from the Visibility drop-down menu if you want other Tinkercad users to see your object and be able to copy and modify it. (Click the link for Attribution if you need more information about making your objects usable by other users and how they can use them.)

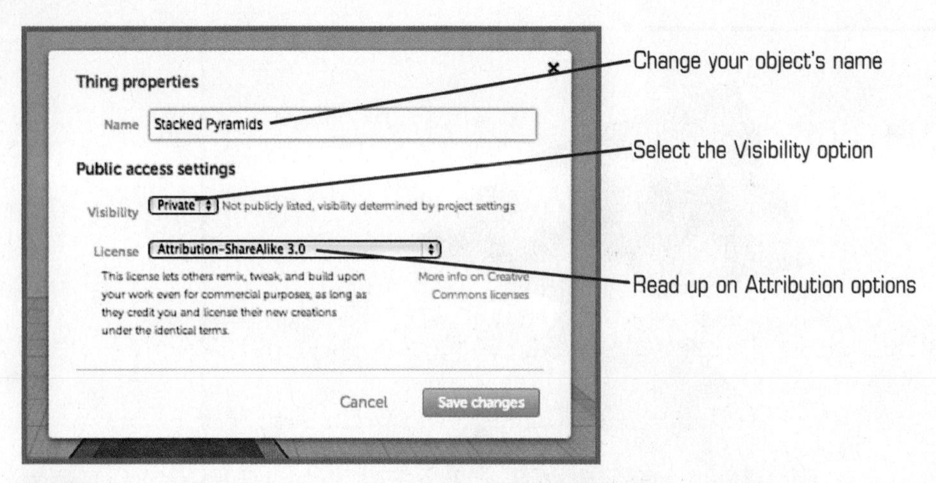

FIGURE 6.27 Save your object and set its viewing properties.

Wrapping Up Tinkercad

With Tinkercad, you can create some amazing objects with a minimum of tools. Don't believe me? Before you exit Tinkercad and move on to Chapter 7, click the Discover button on your Dashboard, and you'll be able to browse hundreds of objects that have been submitted by other Tinkercad users. You'll see a lot of simple objects, but scattered among them are dozens of much more detailed objects. I hope this gives you some ideas, inspiration, and proof that Tinkercad may be the only CAD application you ever need for your 3D printing needs.

Next, I go through the entire process of taking an idea out of my head, turning it into a 3D object with Tinkercad, and then printing it out. Along the way you learn a few more Tinkercad tricks, as well as how to save and export your object to an STL file suitable for printing. Let's go!

Creating a 3D Model with Tinkercad

Chapter 6, "Free 3D Modeling Software," introduced you to Tinkercad, one of the easiest CAD applications available. Add in the fact that it's free to use, and there's really no excuse not to give it a try. If you find it's not for you, you'll want to run through Appendix A, "3D Printer and Modeling Resources," because I'll also be listing some other free CAD applications for you to try (as well as a few good CAD apps that aren't free but won't break the bank).

In this chapter, I use Tinkercad to do some designing and create something that I print on the Simple. I use most of the tools you read about in Chapter 6, but I also show you a few other features and tasks that you can (or should) do when using Tinkercad. I encourage you to log in to Tinkercad and follow along with the chapter.

Hello World

I'm not sure whether you know anything about computer programming, but programmers use what are called programming languages to create the various applications (for mobile devices) that we all enjoy using. For decades there has been a tradition in programming courses and books that has a novice programmer creating a program to put two simple words up on the computer screen: Hello World.

I'm not one to break tradition, so the first item I want to create for my new Printrbot Simple is a variation of the Hello World task. To celebrate the successful assembly and testing of my Simple, I want to print out a small medallion that I can tape or glue to the side of my 3D printer. This medallion will have the words "Hello World."

If you'd like to make one for your Simple, follow along as I show you how I design this small object. Keep in mind that there's not a lot of surface area on the Simple except for the printer bed, and I can't put the medallion there! I've decided that the side of the base shown in Figure 7.1 is a great location, so I've taken some measurements to determine that the medallion will have a maximum length of 50mm, a width of 30mm, and a height of 4mm.

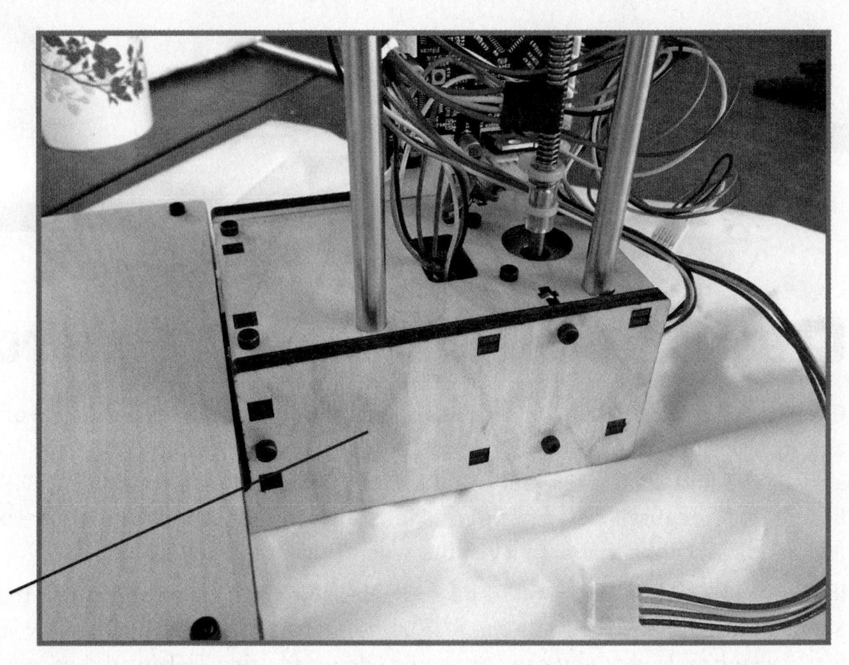

FIGURE 7.1 The side of the Simple makes a great place to mount a medallion.

I'd also like the words "Hello World" to be holes in the medallion so that the wood shows through. I'll be printing with the clear PLA that was provided by Printrbot, but if you have some colored PLA filament, feel free to print it out in color. When I'm happy with the printing, I'll coat it with some primer and then paint it a nice color.

Now that I have an idea of what I want, it's time to open up Tinkercad, log in, and open a new project. As you can see in Figure 7.2, I've created a new project titled Hello World, and it has a blank workspace at the moment.

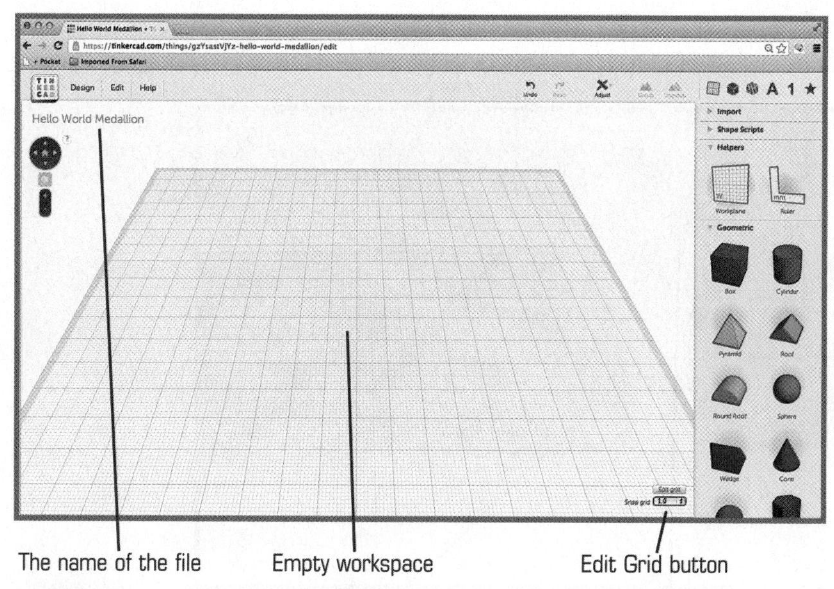

FIGURE 7.2 Open up Tinkercad to begin a new project.

The default size of the workspace is 200mm×200mm, much more than I actually need, so I have plenty of room to work. I start by putting down a Box that I'll resize to a rectangle. Figure 7.3 shows that I've resized the Box object I've dragged into the workspace to make it longer and wider. The dimensions for length and width are visible, but not the height. I've decreased it to a height of 3mm.

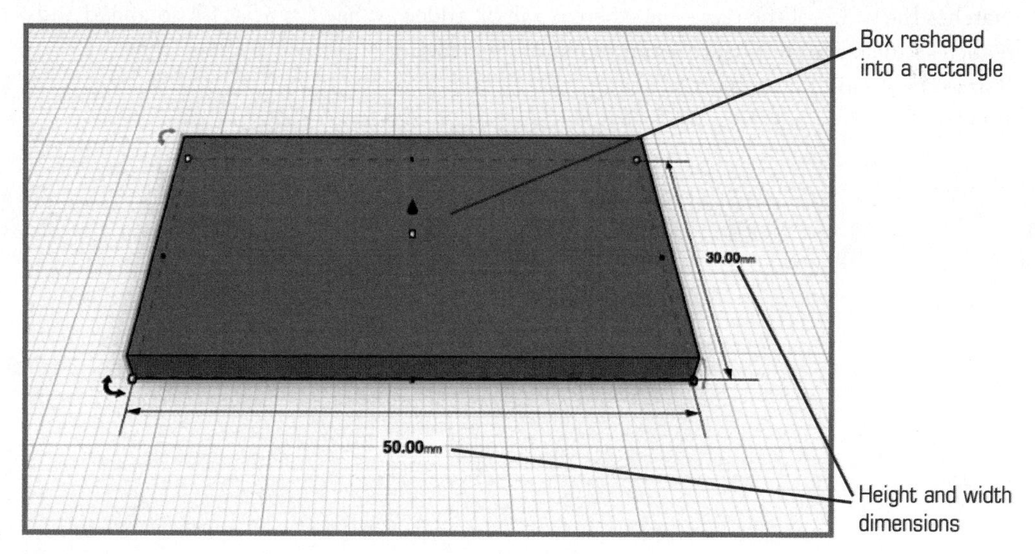

FIGURE 7.3 Start with a rectangle for the basic shape of the medallion.

I'd like to add some rounded ends to the medallion to improve the look. To do this, I'm going to drag in a Round Roof object, as shown in Figure 7.4.

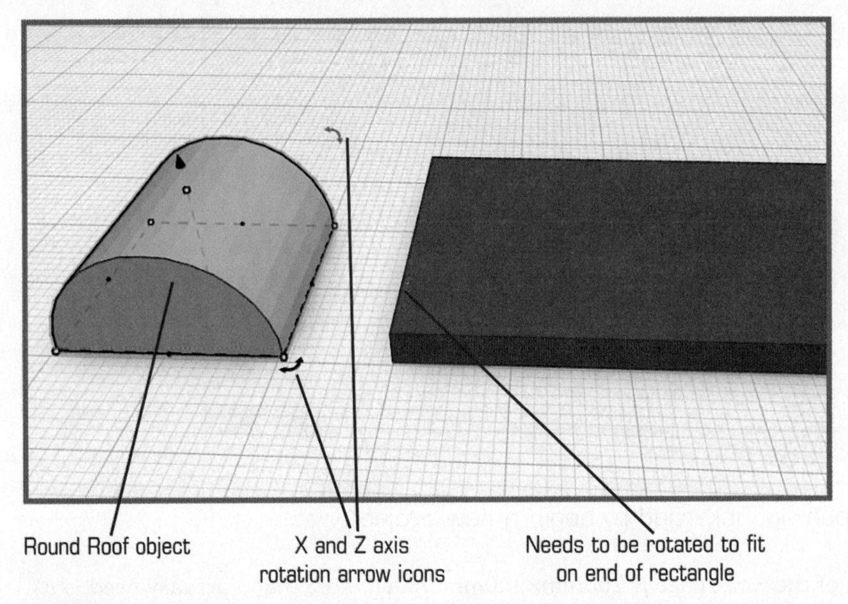

FIGURE 7.4 The Round Roof object will give the medallion curved edges.

If I can rotate this object properly, I should be able to add it to the left edge of the rectangle. But, before I start rotating, it will be useful to resize the object so its diameter matches the width of the rectangle where it will be added. While I'm at it, I'll go ahead and reduce its thickness (height) to 3mm to match the rectangle. Figure 7.5 shows that I've resized the Round Roof object.

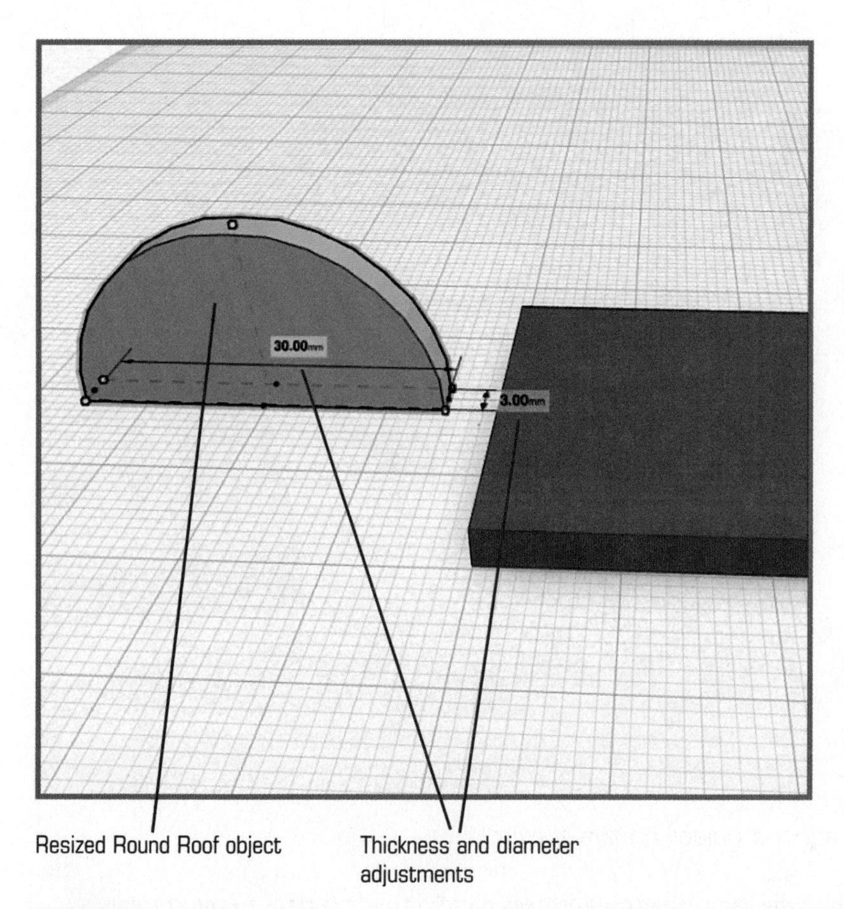

FIGURE 7.5 Resize object to match medallion's thickness and diameter.

Next, I need to rotate the flattened Round Roof object along the X axis. To do this, I move my mouse over the X-axis rotation arrow icon and click and hold as I drag. The angle of the rotation is displayed, as shown in Figure 7.6, and I rotate it to 90 degrees.

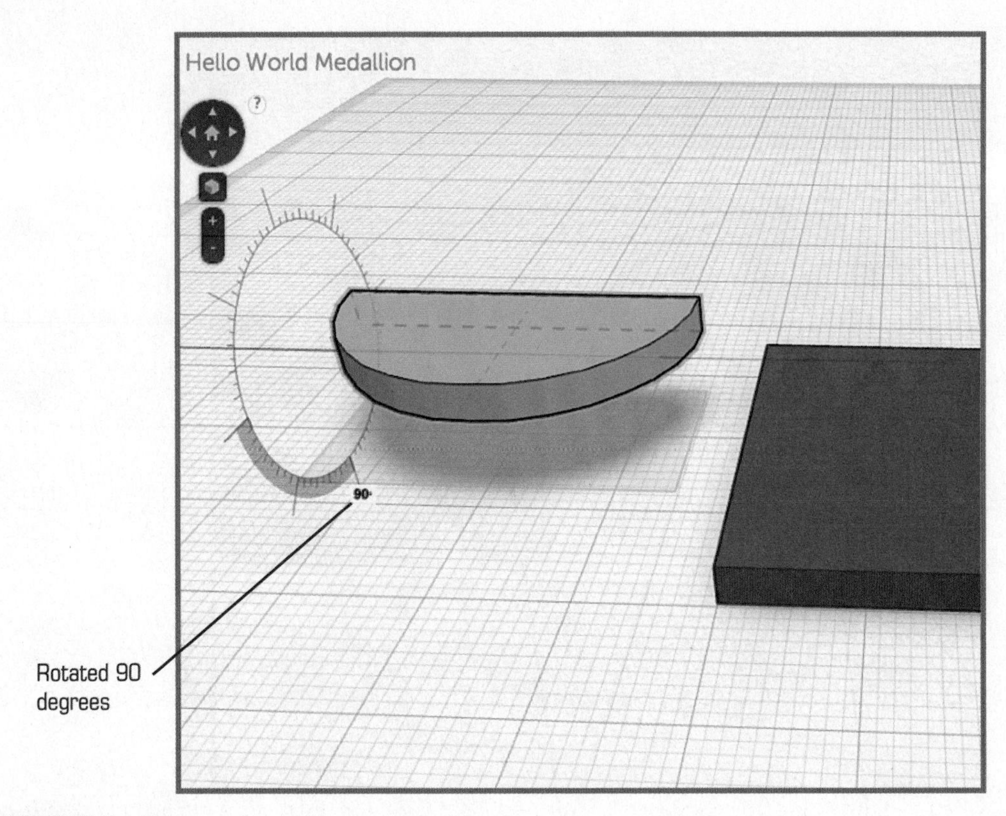

FIGURE 7.6 Rotate the object on the X axis.

Now I need to rotate the same object 90 degrees parallel to the surface of the current workspace (or the Z axis). I'll click the Z-axis rotation arrow icon and rotate the object as shown in Figure 7.7.

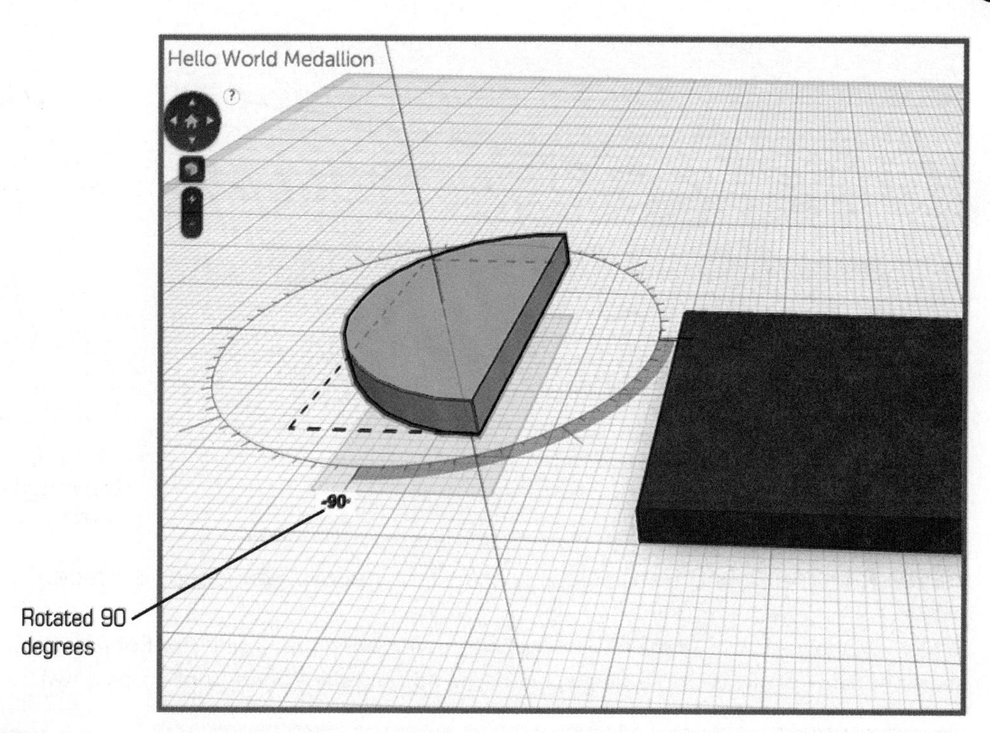

FIGURE 7.7 Rotate the object on the Z axis.

If I rotate the workspace view a bit, I can see that the Round Roof object is floating above the workspace, as shown in Figure 7.8. I need to lower it a bit, and I use the Align feature (refer back to Chapter 6 if you need help) to align the two objects using the workspace that the rectangle is resting on.

FIGURE 7.8 The Round Roof object floats slightly above workspace surface.

Figure 7.9 shows that I use the lower Align dot to first bring the Round Roof object down to share the same plane as the rectangle. I click the dot and the Round Roof drops down.

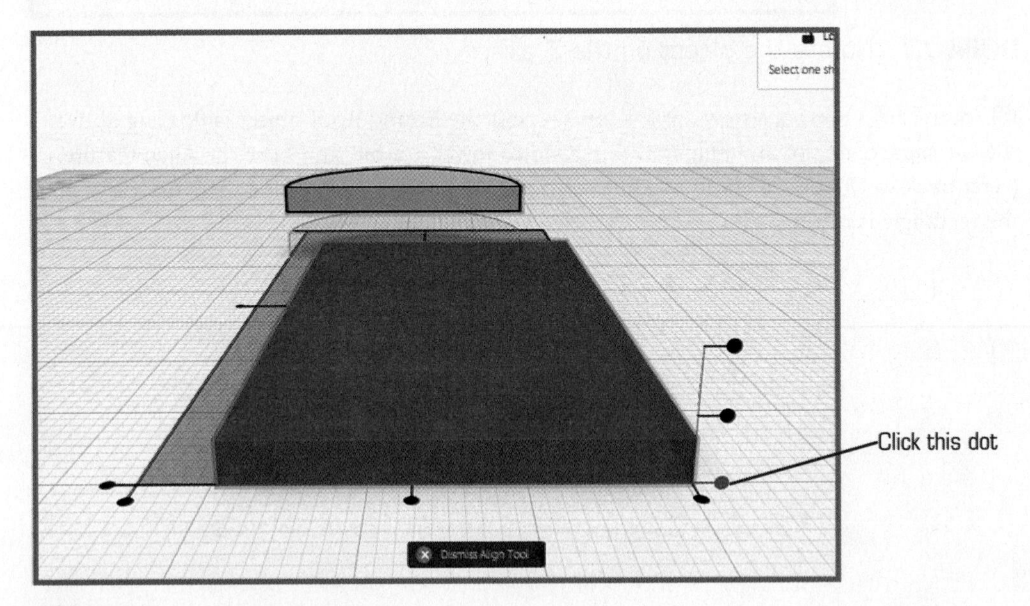

FIGURE 7.9 Round Roof lowers.

Next, I use the Align tool again to force the Round Roof to share the front edge of the rectangle. Figure 7.10 shows that I'm clicking on the lower Align dot so that the Round Roof will move to share that edge.

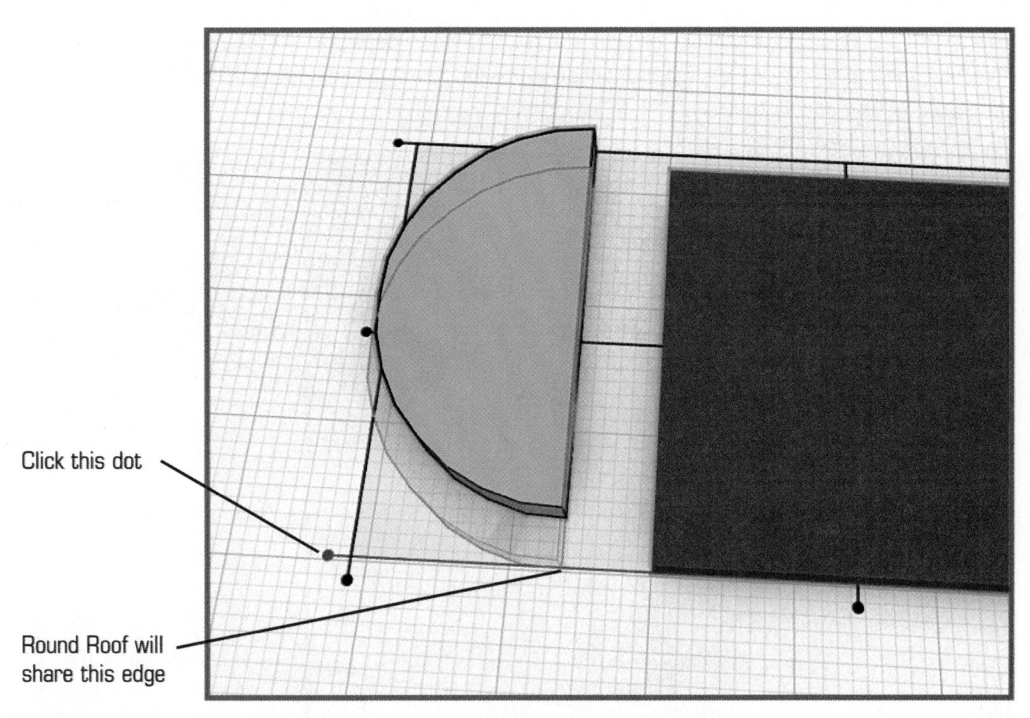

FIGURE 7.10 Align the two objects along a common edge.

Figure 7.11 shows the result. You'll also see a copy of the Round Roof object. I selected just the Round Roof and made a copy of it using Control+C and pasted a new version with Control+V.

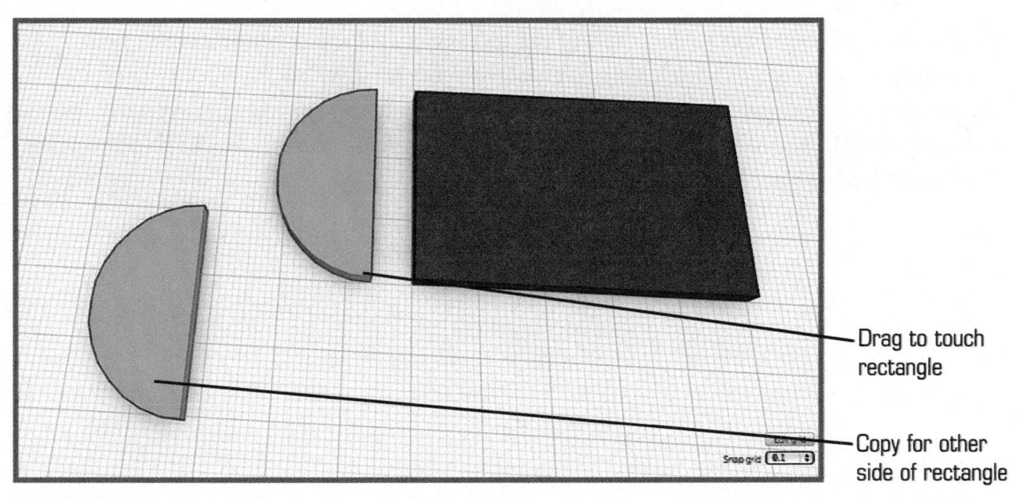

FIGURE 7.11 Make a copy of the Round Roof object for other side of medallion.

You could use the rotate feature to spin the Round Roof copy, but an easier method is to select the copy, click the Adjust drop-down menu, and select Mirror. Click the arrow indicated in Figure 7.12.

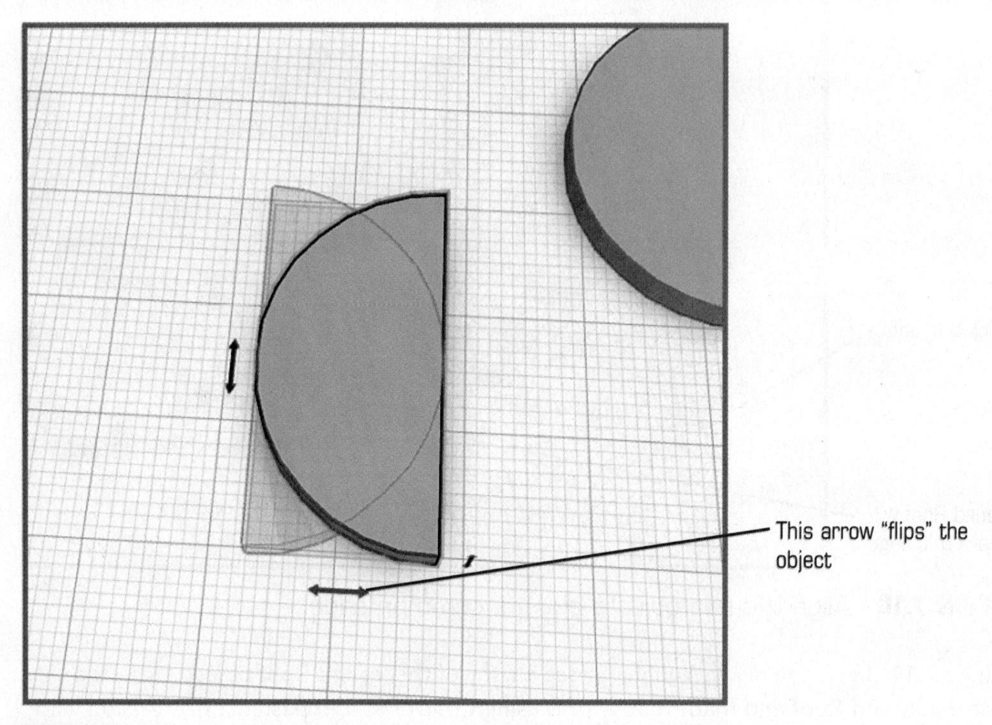

FIGURE 7.12 Flip the object so it is a mirror of the original.

Drag a Round Roof object to each end of the rectangle. Use the Zoom feature to help you make sure the edges are touching. Even better, you can overlap the Round Roof objects with the rectangle by a tiny amount to ensure that no gap exists between the parts. Figure 7.13 shows the outline of the Medallion so far, and you can see that the parts slightly overlap.

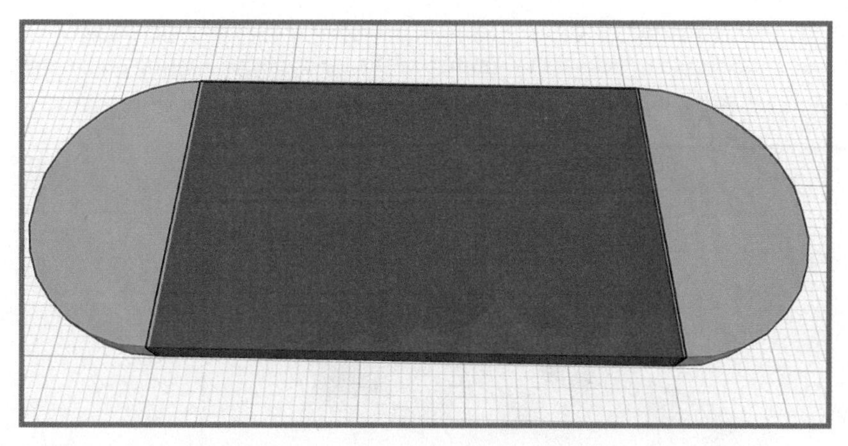

FIGURE 7.13 Slight overlap.

Now I need to group these three objects so they will behave as a single object. This not only makes it easier to drag the medallion around the screen, but it also makes any changes in thickness, for example, apply to all three objects (rectangle and two Round Roof objects) at the same time. To do this, drag a selection rectangle around all three objects and click the Group button indicated in Figure 7.14. Notice that the colors of all objects change to a single color.

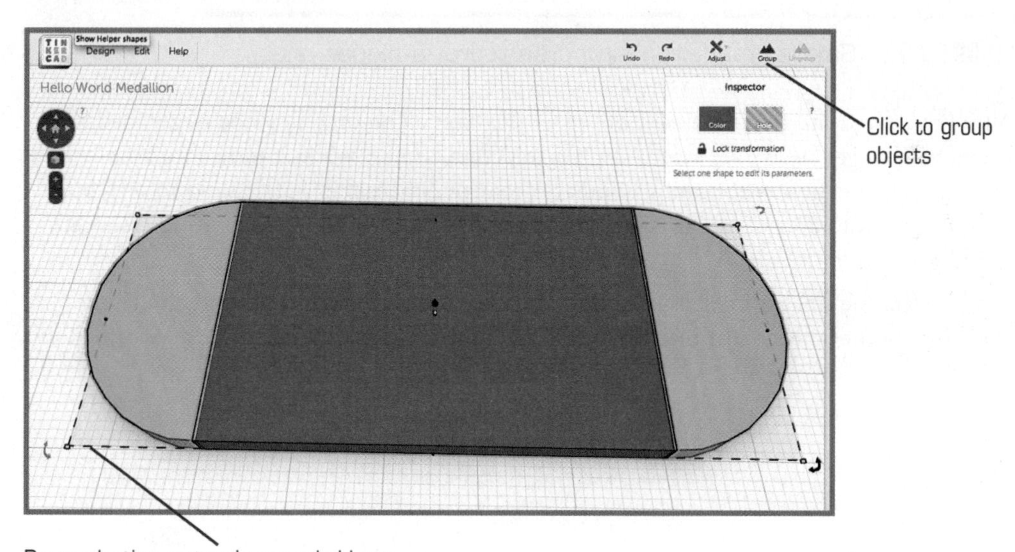

FIGURE 7.14 Group the three objects so they act as one object.

For a decorative element, I'd like to have a slight "lip" that runs around the outer edge of the medallion. To do this, I'll make a copy of the larger medallion, paste it in the workspace, change its height to 1mm, and then place it on top of the larger medallion. You can see this in Figure 7.15. (After reducing the thickness to 1mm, use the black arrow floating above the object to raise it so it's floating at 3mm.)

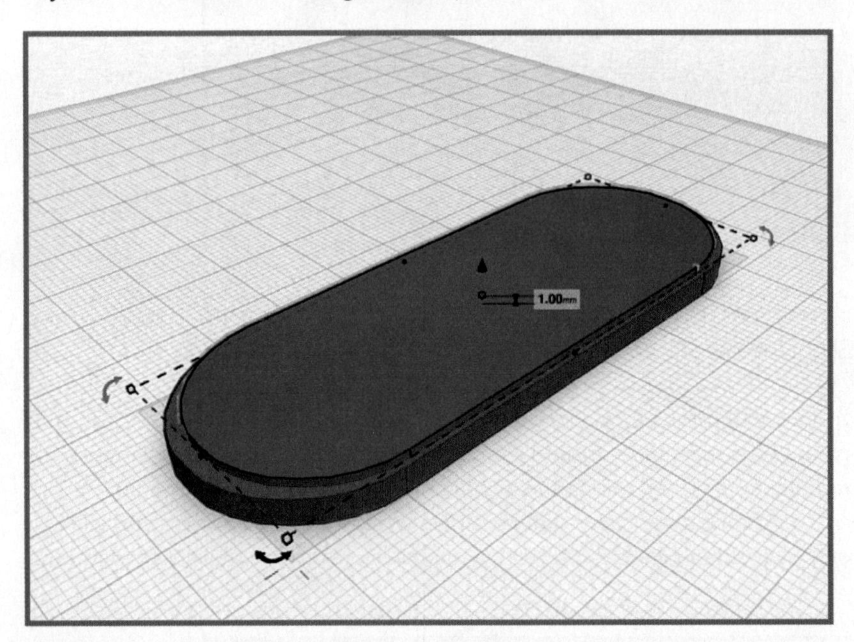

FIGURE 7.15 Smaller medallion sitting on top of a bigger one.

After I'm happy with the placement of the smaller medallion on top of the larger medallion, I want to change the smaller medallion (on top) to a "hole." When it turns into a hole, it will still look like an object I can manipulate (change dimensions, rotate, and so on), but if I overlap it with another solid object, it will remove the overlapped section—in essence, creating a void, or a hole.

First, I select the smaller medallion on top and click the Hole button indicated in Figure 7.16. Notice that the smaller medallion still has edges, but it loses its color. It looks hollow.

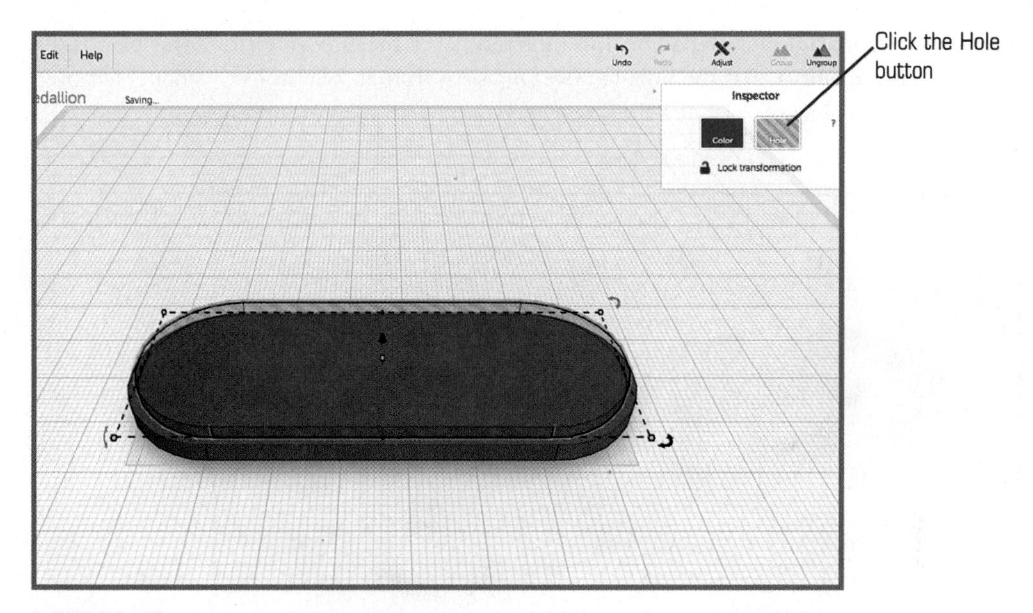

FIGURE 7.16 Change the smaller medallion to function as a hole.

Now all I need to do is "sink" the small medallion (hole) object into the solid larger medallion. I'll sink it only 1mm because that's the thickness of the hole now. Figure 7.17 shows that the hole is now inside the larger medallion.

FIGURE 7.17 The hole is inside the larger medallion.

Next, I select both objects (medallion and hole) and click the Group button. This blends the two objects—remember that the hole removes material from a solid object. What is left over is what you see in Figure 7.18. It's the larger medallion with a lip around a slightly lower inner surface.

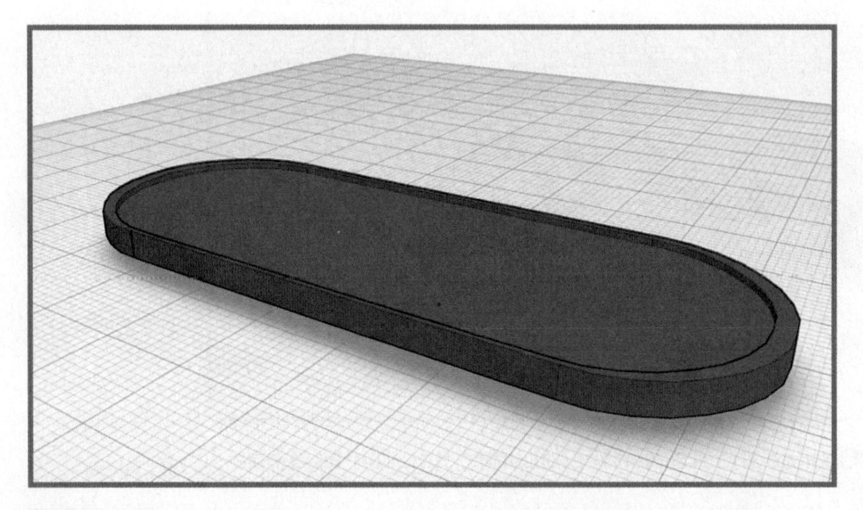

FIGURE 7.18 The hole takes away part of the larger medallion.

Now all that's left is to put in the letters for the Hello World. Fortunately, Tinkercad provides letters (as well as numbers)! It will be easier to first create the Hello World text as a series of letters on a different part of the workspace. I'll need to make them thick enough that they'll go all the way through the surface of the medallion, group them for easier dragging and dropping, and then convert the entire collection of letters into a hole.

Figure 7.19 shows that I've used the Align tool to help put in the letters in a nice, clean looking order. I also increased the height so it's much taller than the thickness of the medallion; this helps me make sure when I convert the letters to holes that they will go all the way through the medallion.

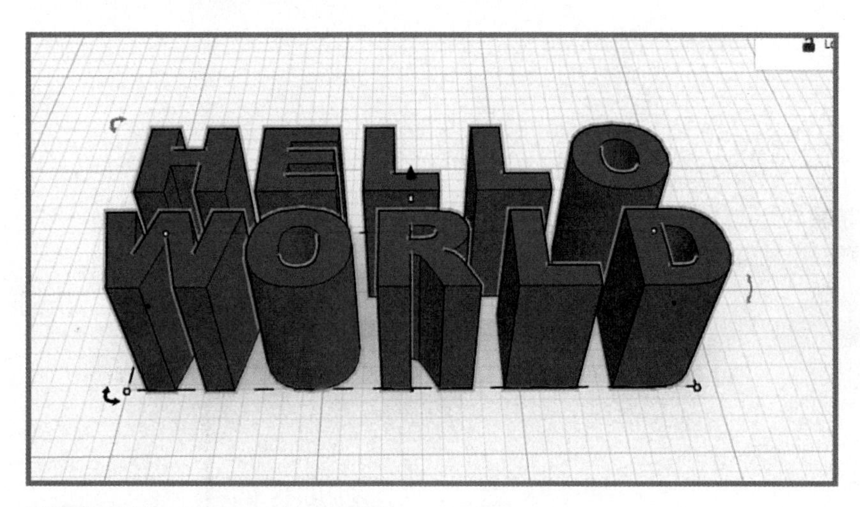

FIGURE 7.19 Grouped letters.

After converting the letters to holes and dragging the group ("Hello World") into the medallion, I drag the group and place it where desired as shown in Figure 7.20.

FIGURE 7.20 Place the letters inside the medallion.

If necessary, I can use the black arrow above the letters to raise or lower the "holes" so they go all the way through the medallion. I've rotated the view to the side so I can see the "holes" piercing the solid medallion, as shown in Figure 7.21.

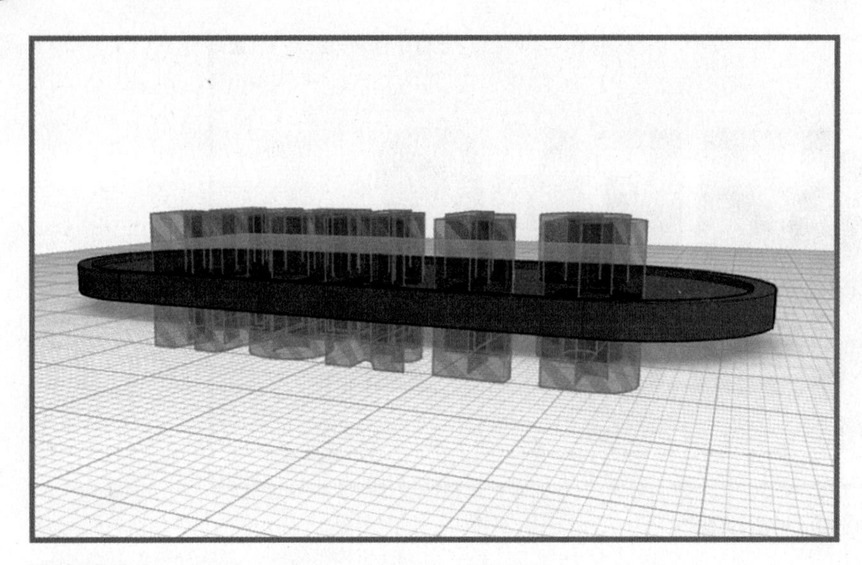

FIGURE 7.21 Holes pierce the entire object.

I'm happy with the placement, so I select everything—medallion and "Hello World" group—and click the Group button to blend it all together. Figure 7.22 shows the final result.

FIGURE 7.22 The final medallion design.

After I'm done with the object, it's time to export it as an STL file that I can use with Repetier to print on my Printrbot Simple. To do this, I click the Design drop-down menu shown in Figure 7.23.

FIGURE 7.23 To print with a 3D printer, you must download the STL file.

A window appears like the one in Figure 7.24. Click the STL button and then choose a location to save the file.

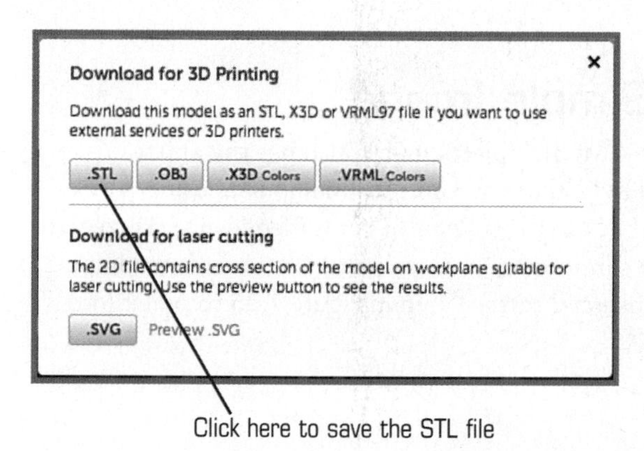

FIGURE 7.24 Click the STL button to begin the download.

Refer back to Chapter 5, "First Print with the Simple," for the steps to import an STL file with Repetier and print your medallion. My final medallion is shown painted and mounted on my Simple in Figure 7.25.

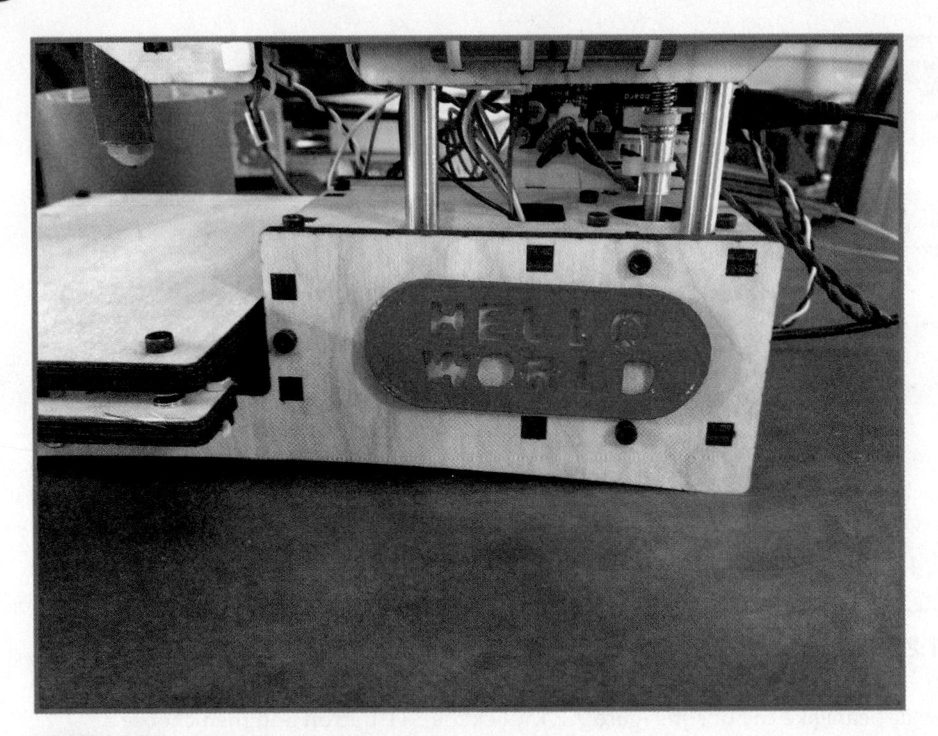

FIGURE 7.25 Medallion painted and mounted

Printing a Sketch or Simple Image

Tinkercad is a great tool for creating your own 3D objects, but it also has the ability to import two specific file types, STL and SVG, that you can then manipulate using the Tinkercad tools. You've already learned about STL, so if you've got a friend who has created a 3D object and has the STL file, she can simply email it to you and you can use the Import button (found on the right side of the Tinkercad screen) seen in Figure 7.26 to pull it into Tinkercad and make any changes you like.

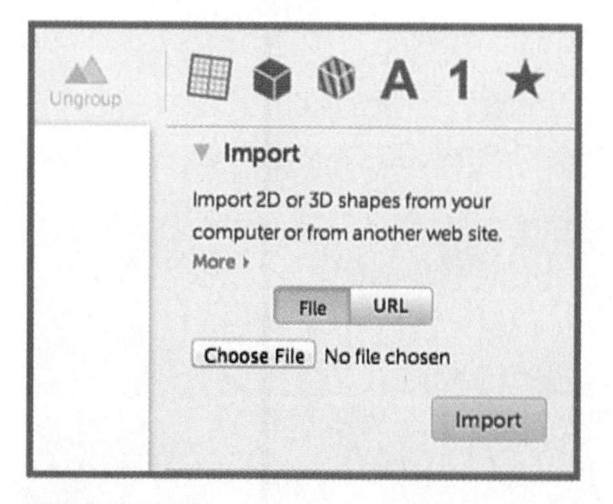

FIGURE 7.26 Use the Import button to pull existing objects into Tinkercad.

But what about that SVG file type? SVG stands for Scalable Vector Graphics and it's simply another format that graphics programs use to save images. You're probably already familiar with JPEG or GIF, for example, but keep in mind that not every graphics program can save as an SVG file type. But if you have one, you can do some really fun things with a 3D printer.

Take a look at Figure 7.27 and you'll see a simple outline of a wizard. Notice that the image is a black object on a white background. You won't need amazing drawing skills to perform this fun task with your 3D printer, but you do need to remember to save your image as an SVG file with the main object being black on a white background.

FIGURE 7.27 Create a drawing and save as an SVG file.
Credits: Brittany Coe, artist; Gamelyn Games, rights owner.

Once you've got your drawing saved as an SVG file, import it into Tinkercad. (You can tweak the Scale and Height before you import the object, but I just import it as is, and then tweak the size and thickness using the Tinkercad tools.) The object will be displayed on the workspace as a flat object like the one seen in Figure 7.28. Increase or decrease the size of the object as desired.

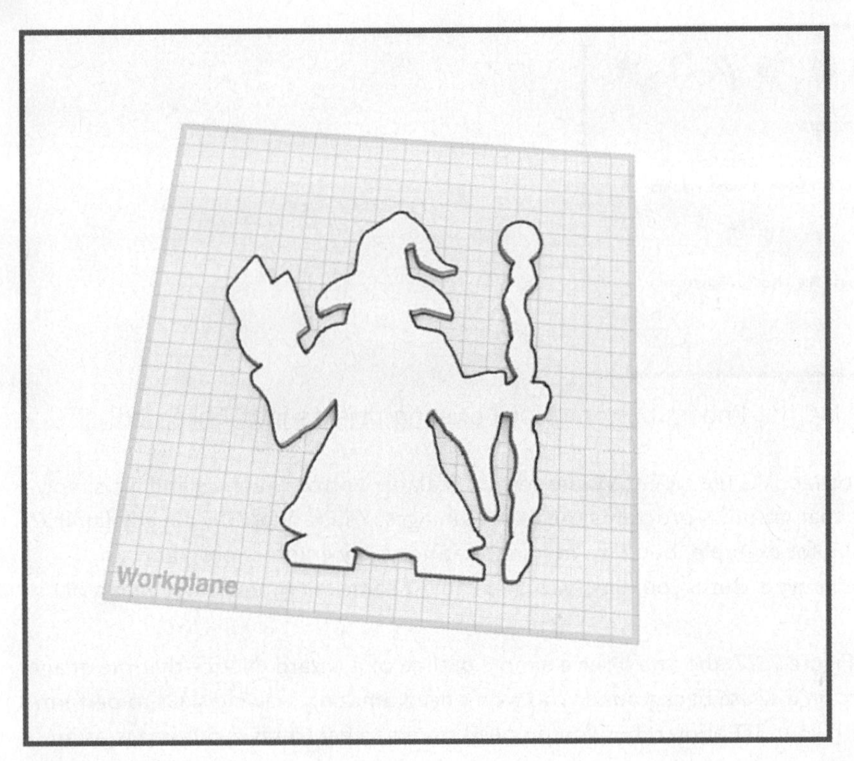

FIGURE 7.28 An imported SVG file is now a Tinkercad object.

At this point, I can increase the height of the object (to increase the printed object's thickness), add new elements, or just save it as an STL file and go straight to printing. I adjusted the size and printed out the wizard that you can see in Figure 7.29.

FIGURE 7.29 The imported object is now a printed object.

Importing an SVG file is a fun way to print three-dimensional objects that you or someone else has drawn; just remember to keep it simple and have the object's color as black on a white background and you're in business. But what do you do if you only have a graphics program (such as Microsoft Paint) that can only save a drawing as a JPEG or PNG file? Fortunately, you can take that image and convert it to an SVG file using a free online tool.

Open a web browser and point it to http://www.online-convert.com and click on the drop-down menu in the Image Converter box shown in Figure 7.30.

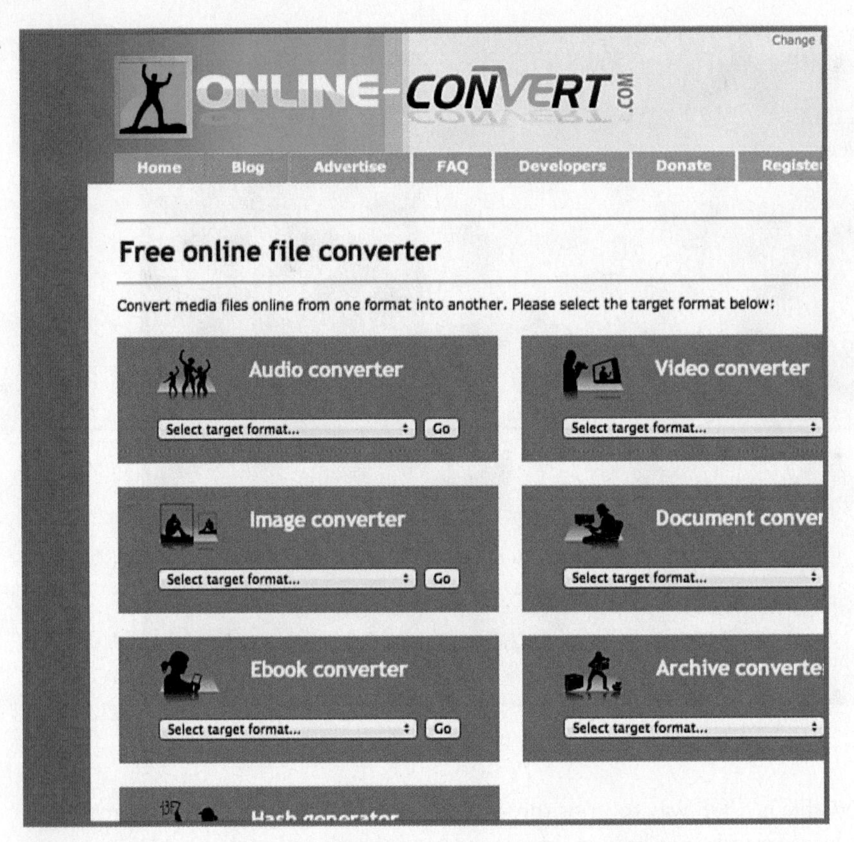

FIGURE 7.30 Use online-convert.com to convert an image to an SVG file.

Select "Convert to SVG" from the many options listed; the website will immediately take you to a control panel like the one shown in Figure 7.31.

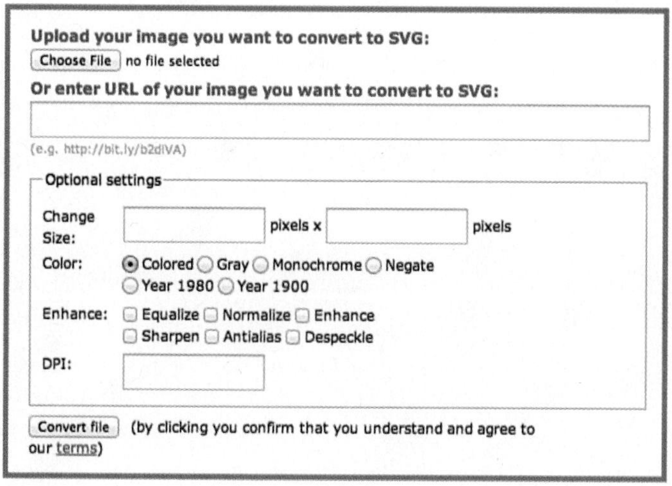

Upload your image you want to convert to SVG:
Choose File no file selected
Or enter URL of your image you want to convert to SVG:
(e.g. http://bit.ly/b2dlVA)
Optional settings
Change Size: pixels x pixels
Color: Colored Gray Monochrome Negate Year 1980 Year 1900
Enhance: Equalize Normalize Enhance Sharpen Antialias Despeckle
DPI:
Convert file (by clicking you confirm that you understand and agree to our terms)

FIGURE 7.31 The imported object is now a printed object.

Click the Choose File button and browse to the file you want to convert. After specifying your file, select the Monochrome option and click the Convert File button and you're done! Your file will automatically begin downloading or you can choose to have the file emailed to you. (If the file doesn't automatically begin downloading, there is a link to download it again. And now you've got an SVG file to import into Tinkercad.

More 3D Modeling Tools

I hope that you found Tinkercad an interesting 3D modeling option. Even though it has a short list of features, you can still do some amazing things with it after you get accustomed to the interface and how the primitive shapes (box, sphere, and so on) can be modified and merged. Given how easy it is to export your models to STL files that your 3D printer can use, you may find yourself using Tinkercad for all your 3D modeling needs.

However, Tinkercad isn't the only game in town. There are plenty of free options as well as CAD applications that you can buy; it really depends on what you need from the software and what you're willing to pay. In this chapter, I introduce you to a family of 3D modeling applications that, as well as other uses, can be used to create 3D models for your 3D printer. I won't be able to cover them in detail, so you'll need to do your own investigating into how to use them. Fortunately, if you've got some hands-on time with Tinkercad, you should find a lot of 3D modeling software interfaces sharing similar features and methods for interacting with models.

Finally, be sure to look over Appendix A, "3D Printer and Modeling Resources," which contains even more references related to 3D printing, including more 3D modeling software developers and their software.

123D Family of Apps

AutoDesk, the owner of Tinkercad, has an entire set of free apps that are perfect for a variety of 3D printing projects. I'm going to cover each of them briefly in this chapter and explain how you might use them side-by-side with your 3D printer.

I selected this family of apps for a number of reasons, but I'd like to share with you a personal one—my 6-year-old son. My son enjoys watching me work with the Simple, but he's not quite skilled enough yet with a mouse to be able to take advantage of Tinkercad's features. He is, however, quite good at using the touchscreen on an iPad, and that's where the 123D family of apps really shines—the apps are easy to use, keeping frustration level low for novices. I think you'll agree with me when you finish reading this chapter. Open a web browser and point it to www.123dapp.com, and you'll see a screen like the one in Figure 8.1.

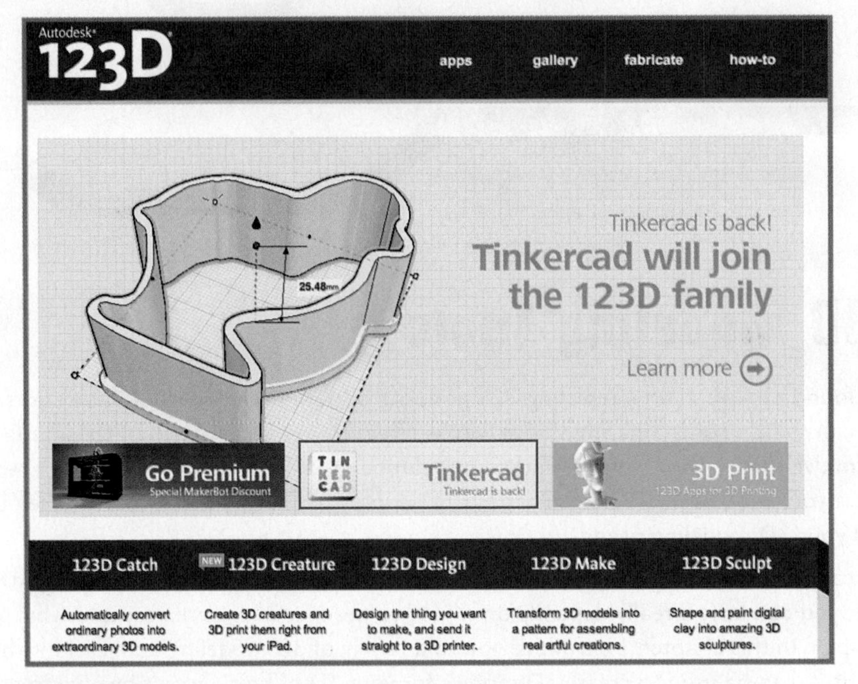

FIGURE 8.1 The 123Dapp home page with the various apps.

There are five apps (six, if you count the newly added Tinkercad that is joining the 123D family) that are all free to download and install on your computer. All you need to do is click one of the app buttons to visit that app's information screen, such as the one in Figure 8.2 for 123D Catch.

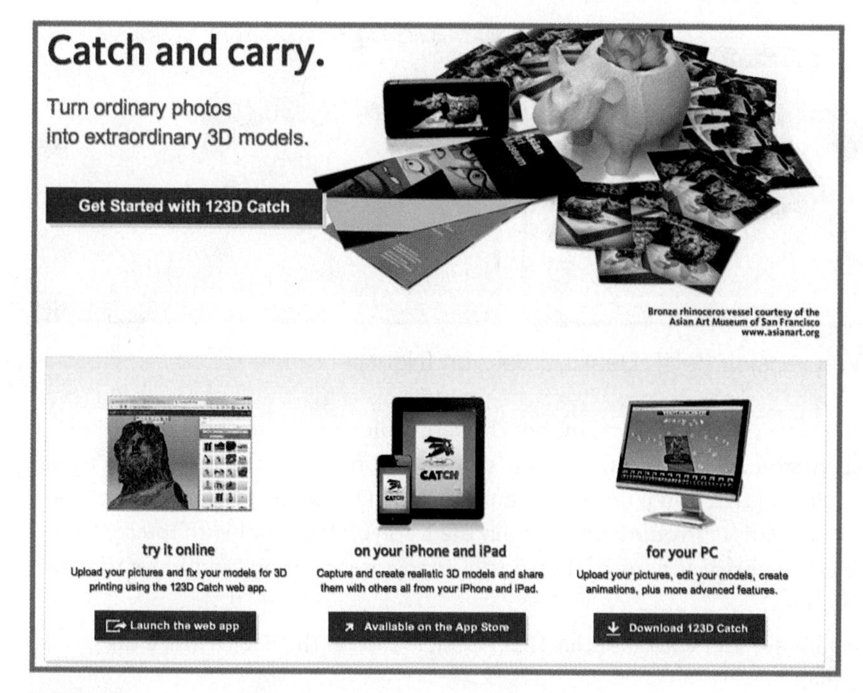

FIGURE 8.2 The 123D Catch app info screen.

As you can see in Figure 8.2, most of the apps (three of the five) come in three formats: an online web-based version, an iPad/iPhone version, and a desktop version. I'm going to cover all five apps but change up the format, so you can see at least one example of each format.

NOTE

As with Tinkercad, the web versions require WebGL, so you'll need to use a web browser (such as Chrome or Firefox) that supports WebGL.

The web versions of the apps are almost identical in look and features as the desktop versions, but I often prefer the desktop versions because they do not require an Internet connection to use. I've also found that sometimes the web versions are a bit sluggish when it comes to dragging and moving objects, and this can be frustrating when you're trying to accurately position an object. Compare the web version of 123D Design on the left in Figure 8.3 to the desktop version on the right.

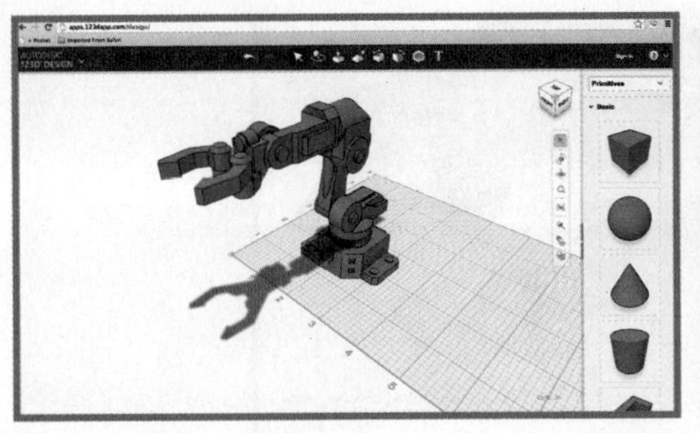

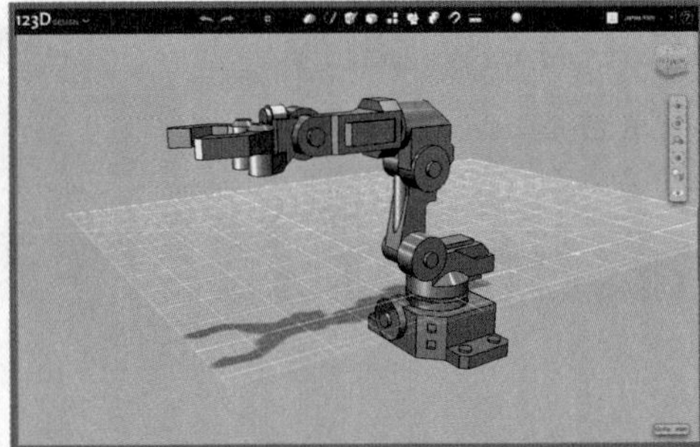

FIGURE 8.3 Web version (left). Desktop version (right).

As for the iPad/iPhone version—I like them, but they are a bit simpler in features than the desktop and web versions. I also find that using your finger on the touchscreen doesn't give you the pinpoint control that you'll need for manipulating 3D models. (Two of the apps are available only on the iPad; I introduce and explain them to you later in this chapter.) That said, my son prefers the iPad versions of the apps over any other option, and I love to use these versions with him.

Take a look at Figure 8.4 and you'll see the 123D Design app on the iPad. Notice the interface is simplified.

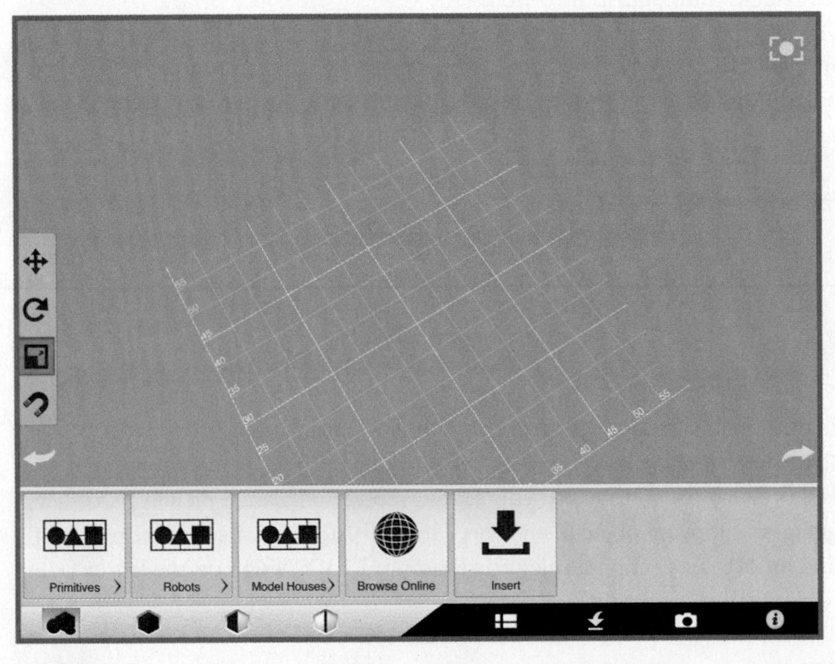

FIGURE 8.4 123D Design on the iPad is simple but has plenty of features.

Ultimately, you should try out all three versions of an app to find the one that works best for you. I prefer the desktop version that lets me use a mouse for fine-tuning my models, but if you like using Tinkercad from within a web browser, you may enjoy using the web versions of the 123D family of apps as well. And if you've got a child wanting to play around with 3D printing, the app versions are going to be a big hit. (I'm not saying an adult can't enjoy the iPad apps, either—you can do some pretty impressive things with them.)

The desktop versions of the 123D apps are available for both Windows and Mac. These apps can be a bit overwhelming at first, but they've also got some great instructional material in the form of videos, a blog, and some built-in tips. You'll also find a forum where you can post questions, learn new tricks, and much more.

I want to introduce you to all five apps, and I'm going to start with the one I use most and the one that is most similar to Tinkercad—123D Design.

123D Design

You can download 123D Design for free by visiting 123dapp.com and clicking the 123D Design button shown back in Figure 8.1. After downloading the desktop version and installing it, open it up and you'll be greeted with the welcome screen.

All the 123D apps open with this welcome screen the first time you open them. They're useful to read over because they give you a brief explanation of the user interface and point out some differences between the apps. If you want to jump right into creating, click the big Start a New Project button, as shown in Figure 8.5.

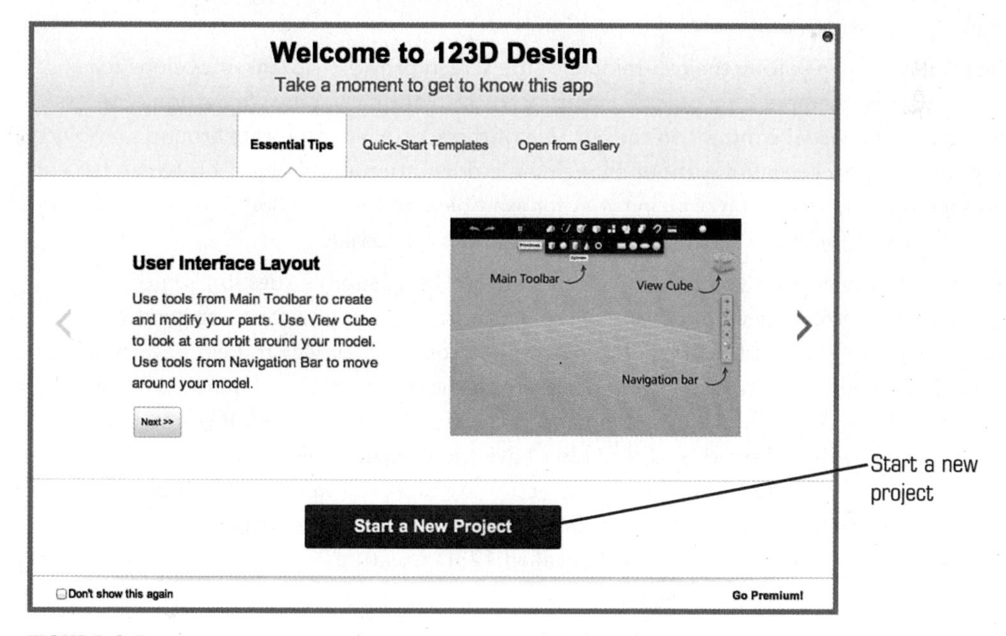

FIGURE 8.5 The welcome screen for 123D Design.

If you're opening up a new project in 123D Design, you'll be greeted with a blank workspace screen, like the one in Figure 8.6.

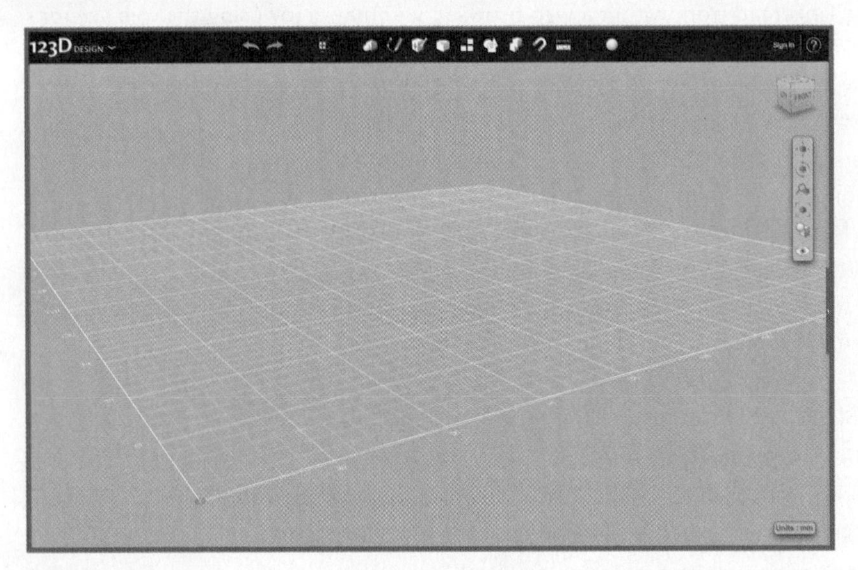

FIGURE 8.6 The workspace screen for 123D Design.

The Design menu lets you save models, export them as STL files, select an online 3D printing service (if you don't own a 3D printer or need a higher quality print than your 3D printer can provide), and more.

The toolbar running along the top-middle of the screen provides dozens of options for adding shapes, manipulating objects, creating groups (or ungrouping), measuring objects, and so on. The Visual controls on the left side of the screen let you rotate around an object, drag the workspace around without changing the orientation of an object on screen (so you can keep the face of an object facing you, for example), zoom in and out, and even turn on and off the capability to view an object's colors and/or materials.

The Help option enables you to view videos, read the blog, submit questions on the forum, and the like. You should create a 123D user account so you can post questions. Fortunately, the forum is full of helpful folks who go out of their way to help novices, so use it often, and your skill level with 123D Design (and the other apps) will quickly improve.

Of the five 123D apps, the 123D Design app is the closest to Tinkercad. If you've been experimenting with Tinkercad, you shouldn't have too difficult a time in creating some very basic 3D models with 123D Design. Still, if you'd like to ease into the 123D apps, there's one very easy (and fun) app to use that can get you a custom 3D model created quickly, if you have access to an iPad or iPhone. It's called 123D Creature.

123D Creature

The 123D Creature app is available only on the iPad, and I love it because I can sit down with my oldest son and we can create monsters together. It's not the most powerful app for creating objects to print with 3D printers, but it definitely gives my son some experience with rotating objects, enlarging and shrinking, and applying colors. While it might seem a bit gimmicky to adults, kids just can't get enough. It's a nice tool for introducing the concept of 3D modeling to a classroom full of kids or just one.

After you download the app from the App Store to your iPad, you begin by posing your monster's skeleton and rotating the arms and legs (or is it tentacles?), changing the thickness and angle of the various body parts. When you're done, you "bake" the skeleton, as shown in Figure 8.7, locking the form so you can begin applying colors and textures.

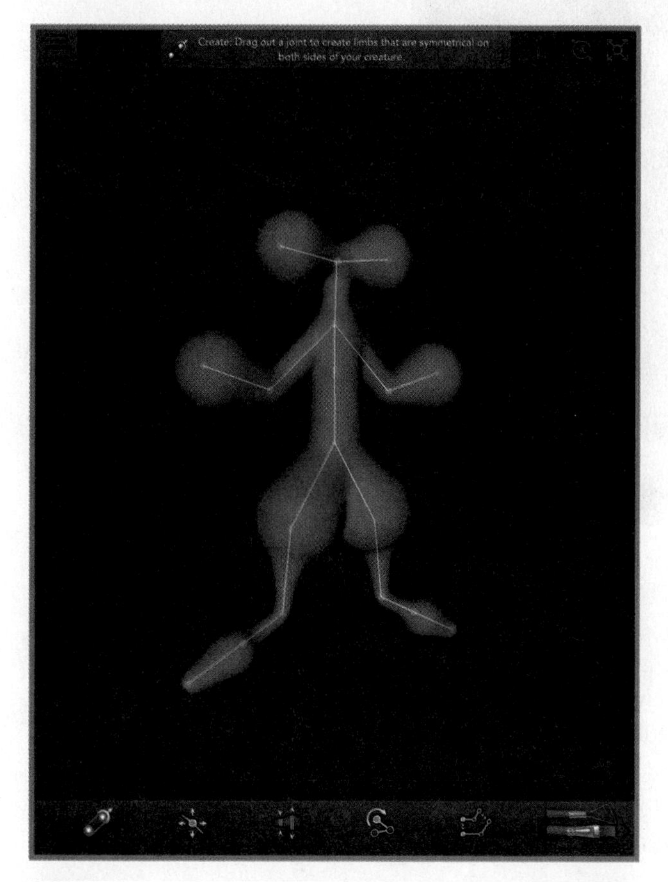

FIGURE 8.7 The "baked" skeleton from the 123D Creature app.

Changes you make to one side of the body are automatically mirrored to the other side—it's fun and crazy. You can sculpt the body, modifying it like you would with your fingers on clay. Apply some color and textures, which are also applied symmetrically, and you're done. My son's crazy looking monster is shown in Figure 8.8.

FIGURE 8.8 Customize your monster with crazy colors and skin textures.

Your creature can be ordered from Sculpteo, an online 3D printing service, but if you want to print it yourself, you should be aware that the 123D Creature app exports the creature only as a Mesh file, a much different format than STL. It can be done, but the steps are much more involved than I can explain here. If you want to convert the Mesh file(s) for your creatures, point a web browser to the following article that explains how to convert Mesh to STL that will be suitable for printing on your 3D printer:

http://www.makerbot.com/blog/2013/03/05/step-by-step-for-123d-creature/

123D Sculpt

Like the 123D Creature app, the 123D Sculpt app is available only on the iPad. (It's possible that AutoDesk may eventually make these apps available for Android devices, but not yet.) And like the Creature app, you'll be using your fingers to mold and shape objects.

As with Creature, it's not the most useful app for adults looking to create advanced 3D models, but it does have its uses. Kids will appreciate the small library of existing objects that can be modified using the touchscreen—no mouse skills required.

You can choose to start with a number of Creature default shapes, such as a head, body, dog, dinosaur, or a few more as shown in Figure 8.9. You can also choose to start with Geometry shapes (cubes, spheres, and so on) or Object shapes that include a shoe, car, jet, and a few more.

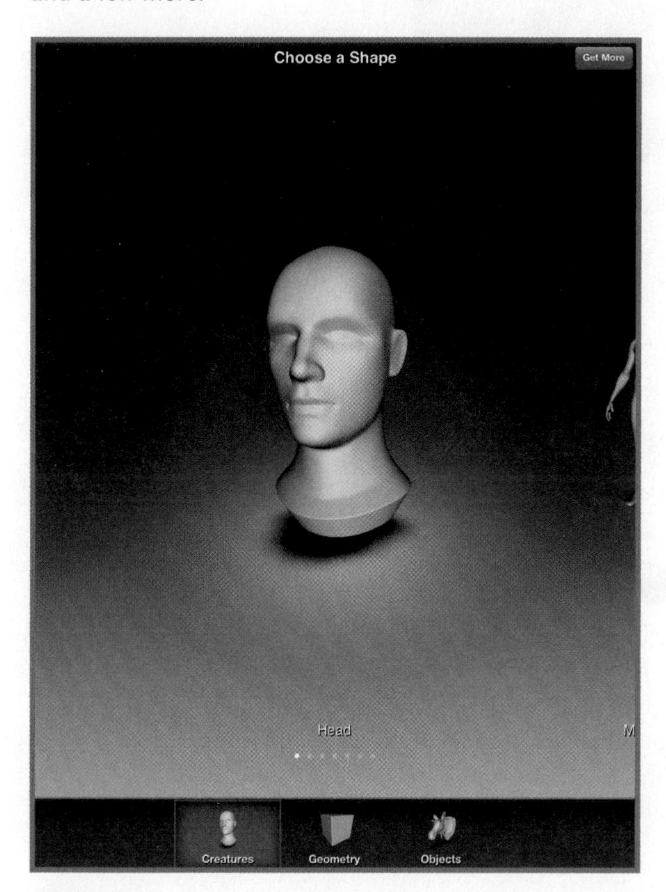

FIGURE 8.9 Tap to start with a head-shaped object.

If you're a serious sculptor, you'll need to invest some time testing and learning the various tools available in the app. But you can also do what my son enjoys—making funny shaped faces and other objects by dragging his fingers through the "clay" and modifying the object onscreen. Figure 8.10 shows his finished car object (and tells me he may not be cut out for actual sculpture).

FIGURE 8.10 The finished car object created in 123D Sculpt.

You then export your final design to the 123D Community, which requires that you create a free account at 123dapp.com, but the file will again be saved in a Mesh format, not an STL file. Follow the link provided in the 123D Creature section previously for details on converting your 123D Sculpt design so it's suitable for printing with your 3D printer.

123D Make

The 123D Make app isn't really a 3D printing app, but if you've spent some time designing a really cool 3D model, you might still want to check it out. In a nutshell, 123D Make takes a 3D model and slices it up into pieces that can be cut from a variety of materials: paper, cardboard, wood, and even metal (if you have the right tools to cut metal).

Figure 8.11 shows a simple 3D model robot that's been "sliced" up horizontally, head to toe; take note of the right side of the screen where the individual patterns are displayed. It's these patterns that you'll transfer to paper or cardboard or wood and then cut out. There are even numbers on them to assist with the proper ordering of the parts.

FIGURE 8.11 3D model sliced up in 123D Make.

123D Catch

The last app in the 123D Family that I'd like to introduce to you is called 123D Catch. This app asks you to first take a series of photographs (from 20 to 70) of an actual object, and then it takes those photos and creates a three dimensional digital version of the object on the screen. You can rotate the object around on the screen, viewing it from any angle. It's an amazing bit of technology, and you absolutely must try it yourself.

The 123D Catch tool is available in all three formats: online tool, iPad app, and desktop application (PC only). As you can see in Figure 8.12, the iPad version lets you view an object on the iPad and rotate it all around.

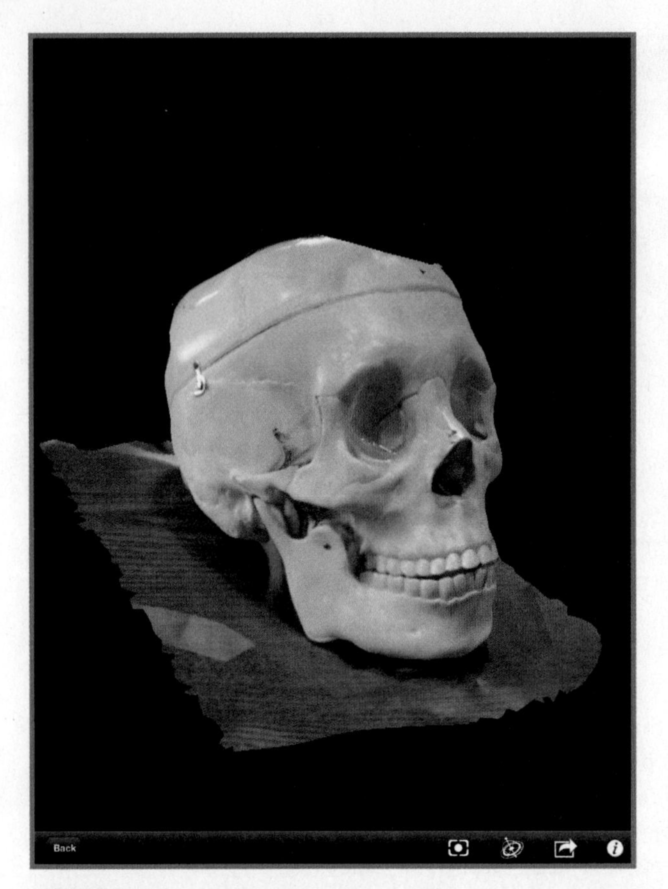

FIGURE 8.12 123D Catch lets you scan an actual object, like this skull, and convert it to a digital file.

The 123D Catch application will also let you zoom in on objects, so you can see even more detail. The trick to creating a detailed 3D object is to take a LOT of photos. The more photos you take (and the closer you can get to the object), the better the final 3D image.

While 123D Catch is designed to allow you to create 3D digital models of actual objects, you might be interested to know that once you've created a 3D model using 123D Catch, you can then print it on a 3D printer. This is a bit more involved, however, so be warned that the steps to do this can be a bit tricky. If you'd like to learn how to use the 123D Catch app to create a 3D model that you can later print, point a web browser to the following online article that offers advice for "cleaning up" the 3D model—basically removing extra stuff like the surface an object is sitting on and other fragments that made it into your set of photos:

http://www.instructables.com/id/Making-a-3D-print-of-a-real-object-using-123D-Catc/

Having Fun

I create a lot of 3D models that never get printed. I just enjoy creating objects that can be created in three dimensions. I like to zoom in and out, adding more details and color. I not only get more practice using the various tools available, but I find that I'm also able to visualize objects better in my mind's eye.

I encourage you to try out all the 123D apps. Not all of them export perfect STL files suitable for 3D printing, but there's always a way if you're interested. (Again, consult Appendix A at the back of the book for some articles that will show you how to do some more advanced 3D modeling tasks, and for some books and websites that you'll find useful.)

Further Explorations

At this point, you should have a fairly strong understanding of what 3D printers and CAD software are and what they can do. There are entry-level 3D printers like the Printrbot Simple, and there are advanced 3DPs like the Replicator 2. On the software side, you've got Tinkercad and the 123D family of apps that provide a lot of creative power to novices, but for users who need more features and tools, there are also dozens of "professional" level CAD applications.

Whether you now consider yourself a beginner or an expert in the field of 3D printing, hopefully this book has given you a glimpse of the possibilities available to you. I mentioned earlier in the book that many 3D printer owners find themselves purchasing or building a second 3D printer—and maybe even a third. It's an addictive hobby. There are always bigger, crazier, and more intricate 3D models to be printed, and that often requires bigger, crazier, and more advanced 3D printers and software.

But even if you're not yet ready to spend more money on a newer 3D printer and software, there are plenty of fun activities for you to consider. If you're a student, a 3D printer can provide some amazing options for assignments. If you're a parent, a 3D printer is a great way to spend family time and learn a new skill together with your kids. If you're a hobbyist, there's the never-ending tweaking and refining of your printing process to get the most perfect model printed.

In this chapter, I introduce some activities that you can do with your 3D printer. Some of these projects go beyond the actual use of a 3D printer into the land of additional hobbies. But that's not a bad thing, is it? Your 3D printer is a tool, and most good tools are simply one part of a bigger (or smaller) project. Following are some projects for your consideration.

Go Bake Some Cookies

You heard me right. I want you to bake some cookies. But these aren't just any old round cookies; these cookies will have a unique shape that you designed. Here's how.

Open a web browser (one that supports WebGL—Chrome or Firefox) and point it to www.thingiverse.com/thing:116042. You'll see a screen like the one in Figure 9.1.

FIGURE 9.1 The Cookie Cutter Customizer link.

Scroll down the page a bit and you'll see the Instructions section. Make certain you've created a Thingiverse user account and are logged in first. After logging in to Thingiverse, click the Open Thingiverse Customizer link. After you click it, the Cookie Cutter Customizer opens, as shown in Figure 9.2.

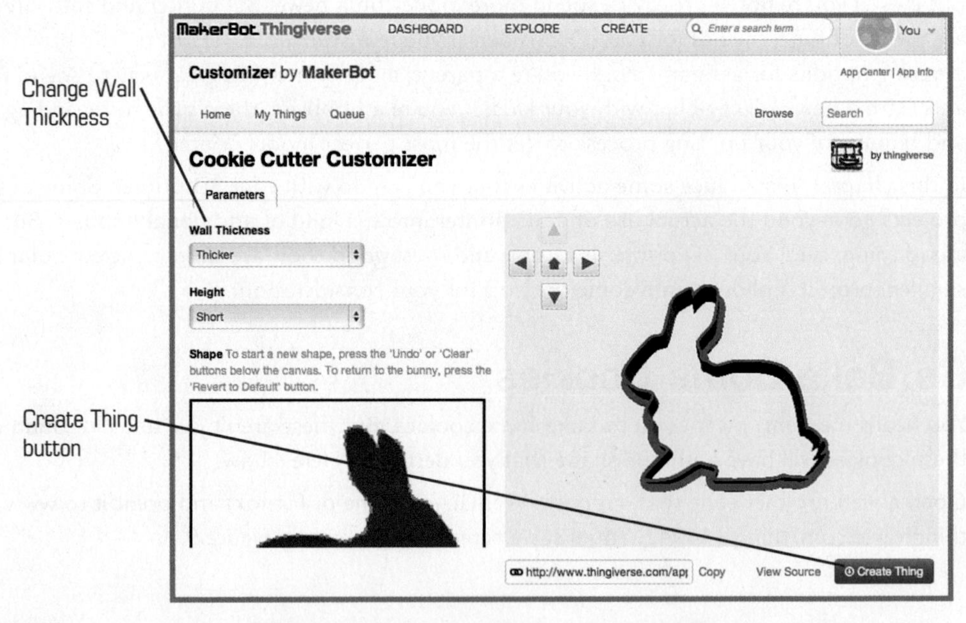

FIGURE 9.2 Modify your Cookie Cutter and export it to an STL file.

Scroll down and click the Clear button to clear out the default image. Draw your cookie cutter shape; the 3D model appears to the right, as shown in Figure 9.3. When you're happy with the design, click the Create Thing button, and then download your STL file, suitable for printing with your 3D printer!

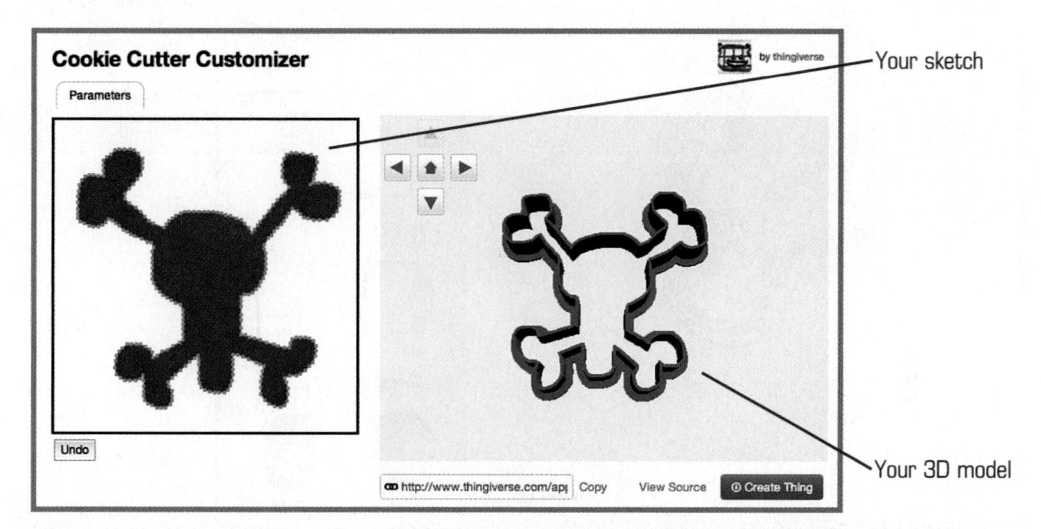

FIGURE 9.3 Draw on left side and your model is created on the right side.

Remember that you can resize your cookie cutter inside a CAD application like Tinkercad. Open up Tinkercad and click the Import option on the right side of the screen. Click the Choose File button and browse to the location of your cookie cutter STL file and open it. The model appears in Tinkercad as shown in Figure 9.4. You might need to reduce its size for it to fit on the workspace; if so, shrink it with the Scale setting.

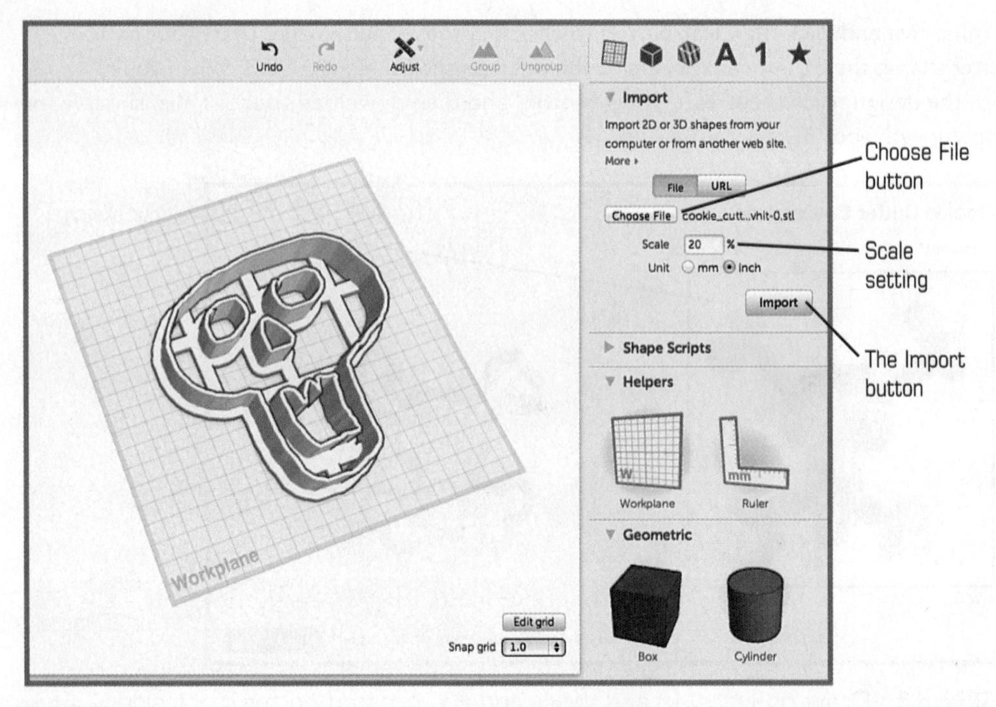

FIGURE 9.4 Import your cookie cutter into Tinkercad to make more modifications.

Don't Forget the Youngest 3D Fans

One of the most amazing things I've discovered with my 3D printer is how interesting it has become to my youngest boy, age 3. The hot end can definitely burn little fingers, so I don't let him get too close, but he loves watching the hot end move and shake and the hum of the motors.

If you've got a 3D printer, I encourage you to share its capabilities with your family, but don't neglect any young children. Unfortunately, 3D printers are not in every school, so if you're open to it, consider reaching out to your nearest elementary school and asking them if they'd like to have a 3D printer demonstration. Don't be surprised if you get a couple dozen requests from teachers in every grade!

And if you're looking for some good ideas for items to print that won't take too long, point your web browser to Thingiverse and search using the keyword **kids**. You'll have to spend some time sifting through all the models, but keep in mind you probably want to print something small. You can always resize an object in your CAD application, but here's a list of some fun little models that I've found that would be perfect as demonstrations of a 3D printer:

- Gekko—http://www.thingiverse.com/thing:28998
- Mr. Jaws—http://www.thingiverse.com/thing:14702

- Mr. Piggy—http://www.thingiverse.com/thing:24669
- Croc Charms—http://www.thingiverse.com/thing:9831
- Space Ships—http://www.thingiverse.com/thing:32141
- Catapult—http://www.thingiverse.com/thing:11910
- Sea Creatures—http://www.thingiverse.com/thing:25276

It's always a good idea to ask a teacher how many minutes you have for your presentation. Find an object and print it in advance so you can make certain it will print in the time allotted. Start your print job the minute you get set up and then begin your discussion. Let the kids ask questions. And take along your test print just in case the object being printed doesn't finish in time. You'll be able to show what the final object looks like. Finally, let the group know at the beginning that the object being printed is going to be given to the teacher or principal—you don't want a large group of kids fighting over who gets the final printed object!

Around the House

After you've gotten comfortable with your 3D printer, it's time to consider putting it to work to make your life a little easier. For example, all kinds of 3D models can provide something useful for your home or apartment. Here's a list of some items you could print with the Simple, but there are plenty more models available on Thingiverse that you can print if you have a 3D printer with a larger print area.

- Cord Pull—www.thingiverse.com/thing:13118
- Tube Squeezer—www.thingiverse.com/thing:1009
- Bag Holder—www.thingiverse.com/thing:26767
- Tie Hanger—www.thingiverse.com/thing:922
- Chip Clip—www.thingiverse.com/thing:4227
- Large Paperclip—www.thingiverse.com/thing:655
- Outlet Cover—www.thingiverse.com/thing:11109

Showing Off

I wanted to include some amazing models that can be printed with a 3D printer, but keep in mind that many of these won't be able to be printed with a Printrbot Simple because of the smaller print bed. If you've got a Simple, you can always scale the model down in size, but when you scale down, the printed model might lose some of the detail (such as a facial expression or text on an object). The only way to know for sure is to give it a try. But if you have a 3D printer with a larger print bed, you might want to try some of these if you're feeling confident in your newfound 3D printing skills.

- Statue of Liberty—www.thingiverse.com/thing:1128
- X-Wing—www.thingiverse.com/thing:98450
- Complex Cube—www.thingiverse.com/thing:123366

- Birdhouse—www.thingiverse.com/thing:115478
- Adjustable Wrench—www.thingiverse.com/thing:87099
- Functional Microscope—www.thingiverse.com/thing:77450
- Simple Scale Mode—www.thingiverse.com/thing:121704

Thingiverse.com isn't the only source for 3D models, however. Following are two other sources, but know that there is some cross-posting; a lot of modelers upload their models on more than one site to maximize the number of visitors who will view them.

- Shapeways—www.shapeways.com/3d_parts_database
- Ponoko—www.ponoko.com/showroom/product-plans/free

OpenSCAD

There's a great tool called OpenSCAD that you might find useful once you get comfortable using your 3D printer and creating 3D models. It's a bit more advanced than typical visual tools like Tinkercad or the 123D apps, but OpenSCAD can be an important tool.

First, open a web browser and point it to www.openscad.org. Before panic sets in, let me explain what you're seeing. You probably saw the word "Programmer" on the screen and got a bit nervous (I know I did), and I will admit that OpenSCAD does create 3D models in a different manner than most 3D modeling applications.

An entire book could be written on the intricacies of OpenSCAD, so I can't go too deep into how to use this tool. But just for a moment, think back to the g-code that you saw created (back in Chapter 6, " Free 3D Modeling Software") that instructs the 3D printer motors on how to operate. That g-code is simply text. Want the hot-end to move up 10mm? Send a g-code command of Z10. Want the hot-end to move 20mm toward the front of the print bed? Issue a g-code command of Y20. It's just code. You get a pretty drawing on the screen of your 3D model, but behind the scenes, all those lines and curves are nothing but simple math equations as shown in Figure 9.5.

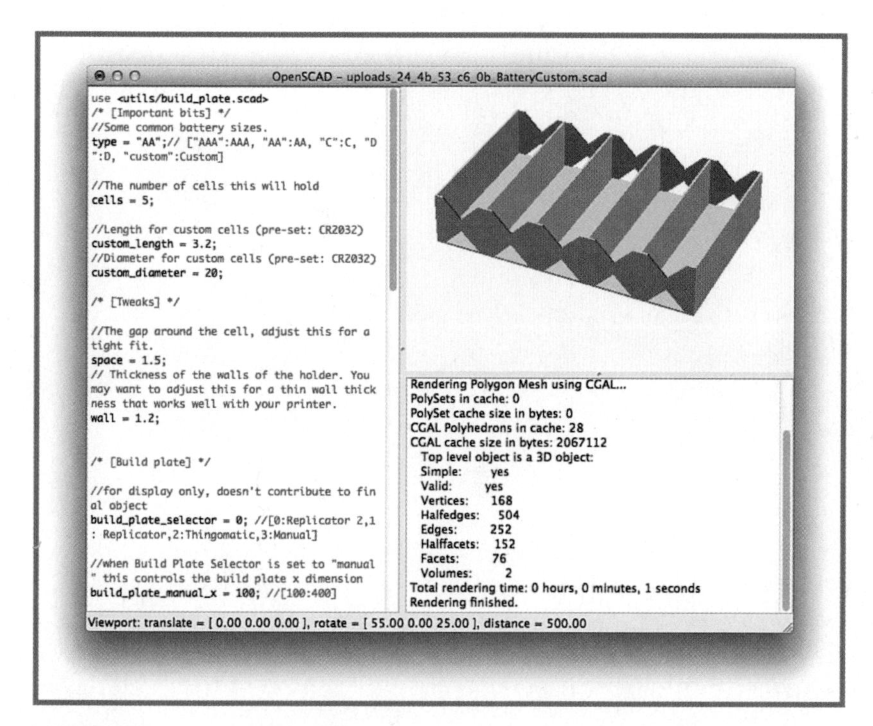

FIGURE 9.5 OpenSCAD uses formulas and programming to create 3D objects.

And that's how OpenSCAD treats 3D modeling. Instead of drawing a cube or a sphere on the screen, you type in the variables that define those objects. A cube, for example, needs to have the length of one of its sides specified. An OpenSCAD command for drawing a cube with the length of each side set to 10mm would simply be **cube(10).** A sphere command of **sphere(5)** would create a sphere with a radius of 5mm. Lines are created by defining starting and end points. Circles and squares (two dimensional) can be specified with **circle(radius)** and **square(size,center)** commands. More complex commands can create curved lines, cylinders, and polyhedrons with the number of sizes specified by you in the code.

One reason a lot of hobbyists enjoy using OpenSCAD is how easy it is to modify an object's properties (such as a square's width) and immediately export a new model. For example, you don't have to go into Tinkercad and drag the side of a square to the new length—just change the value from 5 to 10 (doubling its length) in the OpenSCAD program and you're done.

Of course, OpenSCAD is so much more powerful. Just as Tinkercad allows you to combine cubes and triangles and cylinders to create larger, more complex objects, OpenSCAD works the same way—as you create more advanced objects with OpenSCAD, the actual program becomes more complex. Fortunately the OpenSCAD tool comes with a lot of tutorials and help, so if you'd like to give OpenSCAD a shot, plan on doing some reading first to familiarize yourself with the tool.

Alternatives to the Printrbot Simple

Throughout most of this book, I've referenced the Printrbot Simple and provided photos and screenshots of the various software and objects I've used while tinkering with this inexpensive 3D printer. As of this writing, the Printrbot Simple is the lowest-priced entry into the 3D printing hobby.

However, new products and technologies appear constantly, and prices rarely sit still. With that in mind, I want to close out the book by discussing some 3D printer options that are currently available and some new technologies that are not far off for hobbyists.

One of the great things about the 3D printing hobby is how your skills and knowledge transfer so easily. Should you choose to upgrade to a new, more powerful 3D printer, you'll probably find, as most of us who have been in the hobby for a while have learned, that things just get easier. That learning curve doesn't feel so steep after a while, and you may find yourself looking for new challenges and new technologies to explore.

Let's take a look at where you might go next. You may have started reading this book without a 3D printer, with a Printrbot Simple, or have some other brand and model, so feel free to skip the sections that no longer apply to you. I won't be offended.

Build Your Own 3D Printer

Kits and preassembled 3D printers are great. The out-of-the-box solutions mean you can focus more on the software side and be ready to print immediately (or as soon as you have a model downloaded and your computer connected). The kit solutions save you the time (and the hassle) of having to locate, purchase, ship, and gather all the parts necessary to put together a 3D printer.

But here's the thing: 3D printing has been described as a never-ending addiction, and many 3D printer hobbyists say that they can't seem to stop at owning just one. You may not believe me when I say that, but I have a strong feeling that after you've gotten comfortable with your current 3D printer and all its moving parts, you might start wishing for a larger print bed, or a higher resolution hot end, or maybe even multicolor printing. You'll start developing a mental checklist of all these features and capabilities that your current 3D printer lacks, and you'll want to upgrade or purchase a new 3D printer.

Fortunately, there's always going to be a bigger, better 3D printer available. Unfortunately, that bigger, better 3D printer isn't always in your price range. Or in stock. Or even available in your country (shipping costs for 3D printers overseas can sometimes nearly equal the value of the actual printer itself).

So here's a suggestion: *build your own*. Yes, I'm referring to building your own 3D printer from the ground up—picking the motors, the hot end, the extruder(s), the electronics, and even the frame components. I know it may sound a little intimidating, but if you've made it this far into the book, I hope you have a good idea of just how few parts are required to get a 3D printer working.

A great place to start is with some of the many online forums that cover 3D printing. The grand-daddy 3D printer of them all, the Rep Rap (shown in Figure 10.1), has one of the largest fan bases around, and I can pretty much guarantee that any question you may have has already been asked and answered on the official Rep Rap forum at http://forums.reprap.org. There you'll find thousands of Rep Rap users who have built their own 3D printers along with photos, detailed component lists, and instructions for duplicating their work. You'll also want to take a look at the official wiki page, http://reprap.org/wiki/Main_Page.

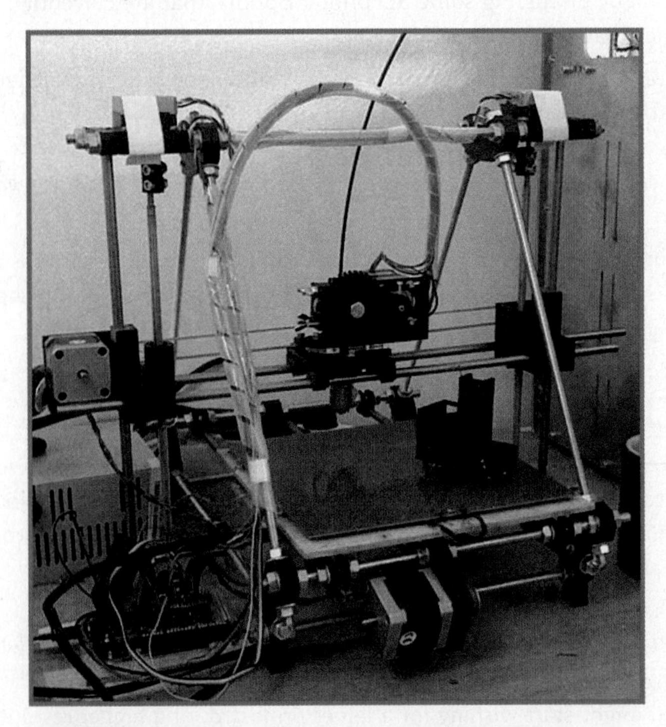

FIGURE 10.1 Just one version of a Rep Rap.

What's great about already owning a 3D printer when you're considering building your own is that you'll also have the capability to print many of the parts that will go into the build-it-yourself (BIY) 3D printer using your existing printer. Gears, connector pieces, and much more can also be printed with your current 3D printer, saving you some money. If your 3D printer doesn't have the capability to print some of the larger parts required, you can always find individuals online who are willing to sell them to you for a small fee.

In addition to the Rep Rap wiki and forum, here are two more places to start your research if you're considering sourcing your own components and building a 3D printer from scratch:

http://www.3ders.org

http://reprapbook.appspot.com

I highly recommend a great product for building your 3D printer's frame—OpenBeam. These aluminum beams (shown in Figure 10.2) are perfect for creating the shell that will hold your 3D printer, and the best part is that the nuts and bolts used to connect all the parts can be purchased at a local hardware store; you won't have proprietary bolts or hard-to-find nuts to hunt down. Even better, visit openbeamusa.com and you can download the files needed to print the plastic connectors. You can buy these connectors from OpenBeam, but why not put your 3D printer to work for you and save some money?

How much money might you save? That's a hard question to answer given just how many variations on a 3D printer you could come up with—size of print bed, number of extruders, type of hot-end, and even size and type of motors. If you purchase a BIY (Build It Yourself) kit from a manufacturer, you'll usually see savings of between $100 and $400, but if you hunt down all the parts on your own, you could feasibly save even more money if you look for sales of the various electronics parts. Frequently manufacturers will have sales on their parts when they are preparing to start selling a newer model—keep an eye on websites that sell individual components (such as printrbot.com, makerbot.com, and others) and you'll frequently see items sold at reduced prices.

NOTE

One word of warning—when buying 3D printer parts individually, be sure to keep track of shipping costs. You might save $10 on a hot-end from Source A and $20 on three motors from Source B. But if you end up paying $15 in shipping costs for each supplier, you might find that the savings disappear quickly. When possible, try to order individual components from a single source, so the parts can be boxed up and shipped in one package.

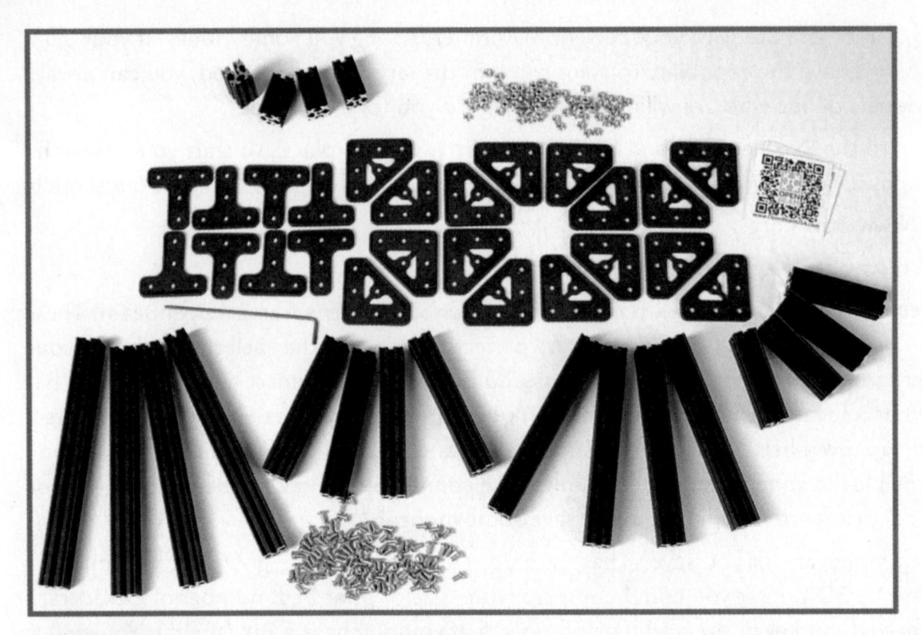

FIGURE 10.2 The OpenBeam aluminum beams are strong, rigid, and easy to connect.

Consider 3D Printer Kits

This section is directed at those 3D printer owners who purchased an out-of-the-box solution. There is absolutely nothing wrong with purchasing a 3D printer that's already assembled and is ready to start printing. Most of the time you can be assured that the printer has been tested before being shipped and that all its parts are oriented correctly, the nuts have been tightened properly, and the electronics work as desired.

Many 3D printer kits allow you to pick and choose from a limited number of options. You may find one seller that offers the same 3D printer with different print bed sizes. Another seller may allow you to select from a couple of extruder options. And yet another seller may offer models in different colors. Printrbot, for example, currently offers four different kits as shown in Figure 10.3. Some of those kits allow you to pick a few customization options, such as selecting between a 1.75mm or 3mm hot end.

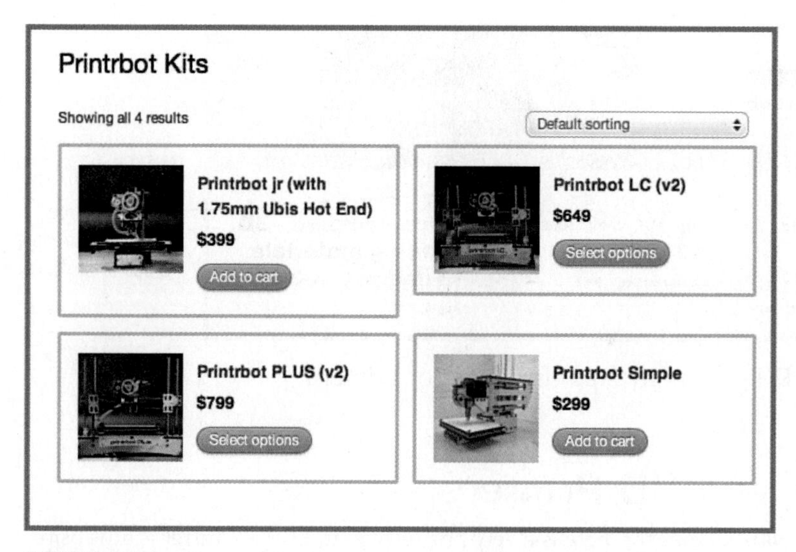

FIGURE 10.3 Four different kits are offered by Printrbot.

Kits are typically priced at a little less than a fully assembled printer, but many sellers will give you a huge price break if you're willing to buy a kit and put it together yourself. I've seen price differences from $100 up to around $400 depending on the brand. Keep in mind that assembling a printer means assigning an employee to build it. And shipping a fully assembled 3D printer is going to (usually) require a much larger box than one where all the parts can be packed flat and in a smaller box. By ordering a kit, you're saving the seller time and money, and a lot of sellers offer some steep discounts to customers who want to build it themselves.

As you read earlier in the book, however, kits should make you pause and do some deeper investigation. You'll want to check out reviews (if they exist) of the 3D printer kits you are considering and see what customers are saying. Are there complaints about missing parts? Do customers mention having to frequently call tech support? What about the 3D printer working when it's all assembled? Although a 3D printer may not be as expensive as a new car, this doesn't mean you should buy the first 3D printer that catches your eye. Get on forums, ask around, and find out which companies are offering great printers and great tech support.

If you're looking for some 3D printer kits, this chapter isn't long enough to list all the various kits for sale. Many companies offer kits. Again, I'm going to point you to http://www.3ders.org. On the website's home page, click the Price Compare tab shown in Figure 10.4, and then click the 3D Printer option that appears. You'll be provided with an updated list containing all the various companies that offer kits (and preassembled printers), along with a link to each company's website. The list is organized by lowest-price 3D printer to highest, so you'll also be able to quickly find companies that offer printers that fall within your price range. Keep in mind that this list is by no means complete; new 3D printers seem to appear almost weekly.

FIGURE 10.4 Click Price Compare to view a list of printers.

Kickstarter and 3D Printers

If you're not familiar with Kickstarter, here is a very brief explanation. Kickstarter is a website that allows individuals and companies to pitch their ideas to the world while asking for financial help. These ideas can be anything from a band needing to raise funds for some studio time to a stay-at-home dad needing financial help to publish his *Cookbook for Kids*. People post proposals for books, music, food, fashion, and more. The pitch is made along with a financial goal. The idea is to woo backers to provide a bit of the money needed to make the goal a reality. It should come as no surprise that many small businesses (and even individuals) are reaching out to Kickstarter to help raise funds to create new 3D printers. Backers can chip in any amount, from as little as $1.00 to $1,000s. The typical reward is an actual 3D printer, fresh from the assembly line, sent to a backer who provides a specific level of funding. The draw of Kickstarter isn't only to be the first on the block to get a company's new 3D printer but also to get it at a slightly lower price than it may be sold at retail.

This is exactly how Printrbot got started. Owner Brook Drumm pitched his idea for the original Printrbot on Kickstarter and asked backers to help him raise $25,000 to produce about 50–100 of his first Printrbot 3D printer. He ended up raising $830,827 in about 30 days (see Figure 10.5.)

FIGURE 10.5 The first Printrbot Kickstarter project.

Over 1,800 backers chipped in the money needed to allow Brook to make his dream a reality. He offered more than a dozen different backer levels, but a minimum pledge of $499 would provide one backer with a Printrbot kit and all the parts and electronics necessary to build it. Higher backer levels offered larger versions of the kit that Brook quickly prototyped and created after seeing the initial demand and hearing backer requests for a larger printer.

Printrbot wasn't the first 3D printer on Kickstarter, but it was one of the most successful. Since then, Kickstarter has continually offered a number of 3D printers and kits to backers. It's one of the best ways to get a new 3D printer at a great price, but it also comes with risks. Kickstarter projects often fail to raise enough funds (backers are charged only if full funding is received) so some projects never get off the ground. Other projects raise the funds but through lack of proper financial oversight and discipline, sometimes run out of funds. There are even a couple of Kickstarter projects where the project owners took the money and ran—this is rare, but it does happen. Do your homework and investigate as best you can any Kickstarter project before you toss in your financial support.

When dealing with Kickstarter, your best bet is to ask questions, weigh the financial risks, and consider the source. I backed the original Printrbot, and I have to tell you that agreeing to pay $500 to someone I'd never met for a product that I'd never seen in person was a big leap of faith. But Brook answered backer questions, provided videos of working printers, and presented himself as someone that backers could trust.

NOTE

I finally did meet Brook in person in April at Maker Faire 2012 in San Mateo, California. He was there to demonstrate his latest printer, the Printrbot Jr and to deliver products to backers (who had backed him in December 2011).

Speaking of Kickstarter and 3D printers, there's another huge 3D printer success story that you should know about—the Form 1. The Form 1 was a 3D printer project looking to raise $100,000 that ultimately raised almost $3 million—$2,945,885 to be exact. The Form 1 is a type of stereolithography printer; instead of melting and extruding plastic, it uses a laser to heat up a special chemical that hardens. It builds up objects in layers just like the Simple, but it's a much more complex (and expensive) type of 3D printer. Still, more than 2,000 backers chose to chip in the funding necessary to make the Form 1 shown in Figure 10.6 a reality.

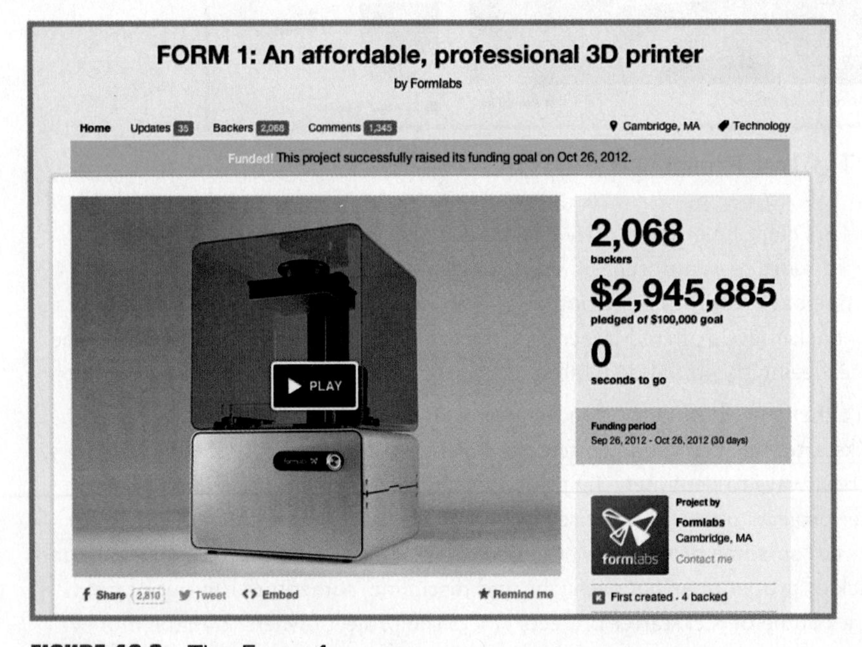

FIGURE 10.6 The Form 1.

The quality of objects printed on the Form 1 far exceeds those objects printed using extruded plastic. But so does the price. Right now, stereolithography 3D printers typically cost between $2,000 and $50,000. Give them a few more years, and they may very well replace the extruded plastic versions of 3D printers. But for now, they're still quite costly as entry-level devices.

Print-It-for-You Services

Hobbyist versions of 3D printers have come a long way in the past few years, and the quality of many 3D objects printed has increased so dramatically that with some higher-end hobbyist 3D printers, it's difficult to tell a printed object from something molded out of plastic. However, this level of quality comes from a more expensive printer—something not available to most homes or schools. So what do you do if you find yourself needing to print out an item in plastic but need a higher level of print quality than what your current 3D printer offers and you can't afford (or don't want) a higher-priced 3D printer? Easy. You outsource it.

A number of 3D printing services will take your 3D model files and print them for you at a much higher resolution and quality, and not just in plastic. Metals, ceramics, and many other materials are available if you've got the funds. Given the high cost of these specialty 3D printers, you may very well find that having your object printed by a third party is an ideal solution.

Two of the biggest names in 3D printing services are Shapeways and Sculpteo. Both allow users to upload their digital 3D model files and then select the material, size, color, and print quality, as shown in Figure 10.7. Provide your credit card number and in a few days to weeks, you'll be holding your submitted 3D object in your hands.

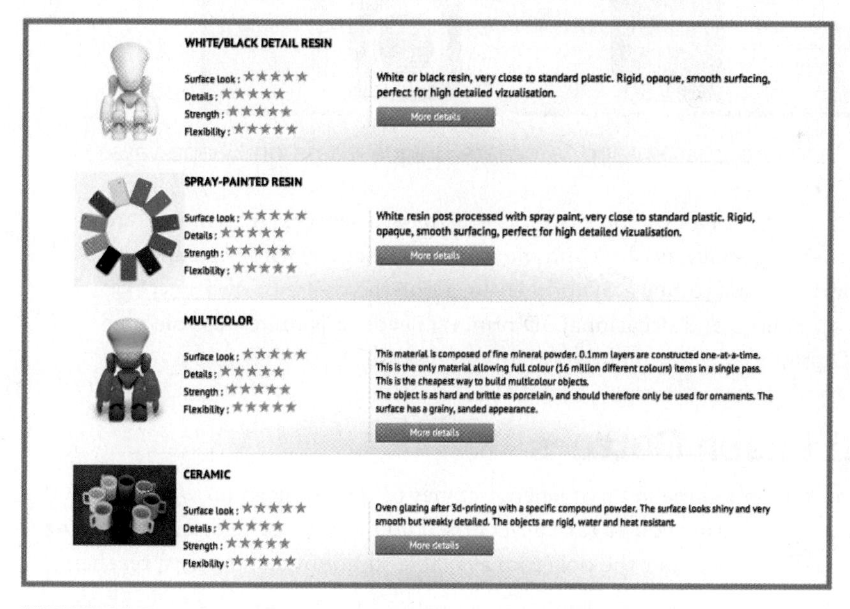

FIGURE 10.7 Sculpteo lets you select materials and colors.

Both Sculpteo (sculpteo.com) and Shapeways (shapeways.com) offer users basic tools to create custom 3D objects. Sculpteo, for example, has a very simple 3D modeling CAD application that runs right from your web browser and lets you create things like keychains,

3D letters, or even a 3D object from a photo or image. Shapeways offers access to a number of custom creation apps, such as a ring designer or a custom poker chip creation app, as shown in Figure 10.8.

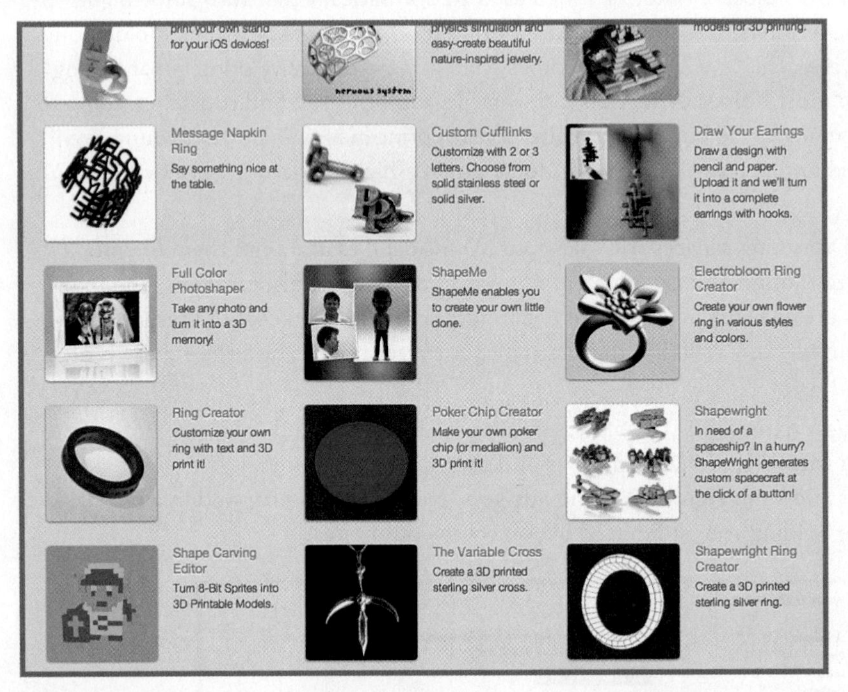

FIGURE 10.8 Choose a custom app to create unique items on Shapeways.

Shapeways and Sculpteo aren't the only two print-it-for-you services. Many more are popping up all over, especially as more and more people discover 3D design and printing. Not everyone has the funds to buy a 3D printer, let alone the desire to own one. For someone with very limited and occasional 3D printing needs, it is quite possible that investing in a 3D printer may not make good financial sense.

CNC and Laser Cutters

3D printers aren't the only game in town when it comes to designing your own things. A 3D printer is great for allowing you to take a 3D object you've created in software and then make it real, but a 3D printer isn't the only tool available to hobbyists, parents, teachers, and students.

Consider that if you've learned your way around any of the number of CAD design applications out there (such as Tinkercad or any of AutoDesk's tools), you've also opened the door to other useful "maker tools," namely a CNC machine and a laser cutter. You may not be familiar with these tools, but you'll be happy to know that learning to use them will be considerably easier if you've gotten comfortable with CAD and using a 3D printer.

Entire books have been written about CNC machines, and the reason for the lack of books on laser cutters is probably because of their high price. I don't have room in this chapter to go over every technical detail about CNC and laser cutters, so I'll explain them relative to the 3D printer knowledge you now possess.

A 3D printer creates objects by building them up in layers. This is called an additive process because something (plastic, for example) is being added over and over to create a three-dimensional object.

Now compare this to the subtractive process of a CNC machine. CNC stands for Computer Numerical Control machine. Just as you use a computer to control the motors of a 3D printer to deposit molten plastic in specific spots, a CNC machine uses a computer to control motors that move a milling bit around to cut away material at specific spots.

NOTE

A CNC machine is all about taking away, not adding. A 3D printer places molten plastic (or other material) onto the work surface to create an object. A CNC machine takes away material (such as wood) so that what's left behind becomes the desired object.

Take wood, for example. If you've seen a master carver use a knife to whittle away bits and pieces of wood to create a sculpture, you're getting the basic idea of a CNC machine. Because a computer is capable of much finer movements (using special motors, like servos), with the right milling bit, amazingly complex and detailed items can be created in wood, plastic, metal, and dozens more materials (depending on the strength of the milling bit and other factors). Take a look at Figure 10.9 and you can see an example of the kinds of objects that can be made using a popular CNC machine called a ShopBot (www.shopbottools.com).

FIGURE 10.9 Objects created in wood by a CNC machine.

CNC machines come in all shapes and sizes, from small, shoebox-size milling machines to garbage-truck-size machines. Some of them operate only in the three basic axes (X, Y, and Z), whereas others bring additional features, such as the capability to cut while the object spins (called the fourth axis) like a lathe.

And what about a laser cutter? The name should give you some very obvious clues about what it does. It uses a high-wattage laser beam to cut out objects. This work is done in only two dimensions (X and Y). Although creating flat objects to be cut out is much easier than designing a 3D object, laser cutters are still considerably more dangerous to operate than a 3D printer. Laser cutters can damage eyes and skin if not used properly, and there's also always the risk of fire because of the high heat of the laser beam being used on raw material.

Again, I mention CNC machines and laser cutters only to give you additional tools to research and discover. Just as there are websites and forums devoted to 3D printing, you'll find matching resources available for CNC and laser cutters.

NOTE

Want another nonstandard tool for your workshop? (By nonstandard, I mean tools that don't include standard stuff like a drill press, table saw, router, and the like). In addition to a 3D printer, laser cutter, and CNC machine, you might also find a plasma cutter useful. It's much more of a fire hazard, but it can cut through thick metals like butter. It operates similar to a laser cutter and cuts shapes from sheet metal. As with any tool, do your research and know the risks and dangers that come from operating that tool.

Scanning Objects

You learned earlier in this book how to create your own 3D models using CAD software such as Tinkercad, but you'll be happy to know that you don't always have to start from scratch. Enter the 3D Scanner.

Just as a scanner can scan a printed photo or document and create a digital version that is stored as a file on your computer, a 3D scanner can scan a 3D object and convert it into a file that can work with a 3D printer.

Take a look at Figure 10.10 and you'll see one of the newest products from MakerBot called the MakerBot Digitizer.

FIGURE 10.10 The Makerbot Digitizer 3D Scanner.

In a nutshell, you place an object on the front plate. This plate rotates and a special camera on the Digitizer takes digital pictures of the object. These images are then assembled using special software that creates a 3D model suitable for printing. The Digitizer is designed to connect directly to a MakerBot Replicator 2 so that you can scan an object and send it directly to the Replicator 2 to be printed in plastic. But the Digitizer can also save your scanned object to a file that can be used with the 3D printer of your choice.

Obviously the size of the object you can scan is limited to an object that can fit inside a circle with a diameter of 8 inches (approximately 20cm) and a maximum height of 8 inches. (If you've got kids, imagine being able to scan in one of those crazy-complex LEGO pieces and being able to scan it in and print out as many as you'd like!)

Just as 3D printers continue to drop in price and become easier to use, so will 3D scanners. The day is fast approaching where you'll be able to scan in larger objects and create as many as you need—imagine printing special chairs for party guests and then recycling them when you're done. Don't laugh—one day you'll be printing all sorts of objects either from scans or downloaded files. The future of 3D printing (and scanning) is really just getting started, and it's going to be a fun ride seeing what comes next.

Where Do I Go From Here?

So, what do you think about 3D printing? It's an amazing technology, and it's only in its infancy. Imagine where 3D printing technology will be in 5 years...10 years...50 years!

But right now, you have the opportunity to have some real fun and maybe even design some useful items with a 3D printer. It's not futuristic technology—it's here today, and you're ready to go.

If you're like me, you may have been bitten by what they call the maker bug. It's a deep desire to tinker, to make things, to fix things, and to try to express your creativity by adding new tools and skills. 3D printing is certainly a fun technology, but it's actually quite useful; many companies use 3D printing to create prototypes of products that will eventually be sold in stores. Inventors use 3D printing to test out sizes and shapes of items, such as gears and cases. Teachers use 3D printing as a tool for introducing the power of CAD software that is the core of many technologies and jobs of the future.

But 3D printing isn't the only technology that can satisfy that maker bug. Hobbyists and professionals around the world have access today to some amazingly powerful tools that weren't readily available just 10 years ago. Costs have dropped, sizes have decreased, and tools that were previously available only to multimillion-dollar manufacturing corporations are now being found in garages, workshops, and schools everywhere.

I'm not talking about the standard shop tools, like table saws, drill presses, routers, and lathes. I'm talking about more powerful tools, and I introduce you to a few of them as I wrap up this book. You should have a good grasp of how a 3D printer works and the hardware and software used to make the magic happen. That knowledge is going to help you make another jump (if you want to do so) to some other tools that will be much easier to learn about now that you're experienced with 3D printing.

CNC Machine

The first tool I want to discuss is called a CNC machine. Before I show you what one looks like (and many different shapes and sizes exist, just like 3D printers), I'll go over briefly what a CNC machine is and what it can do for you. I'm going to use what you know about 3D printers to compare and contrast with a CNC machine.

First, a 3D printer performs what is called an *additive process*. Additive—something is being added. Pretty straightforward. What's being added is melted plastic, layer by layer, to create a solid object.

A CNC machine performs what is called a *subtractive process*. It takes something away. And that something depends on the material you are working with. A CNC machine uses a milling bit in lieu of the hot end found on a 3D printer. The milling bit looks like a drill bit, and it spins at a very high speed. If you can imagine a 3D printer with a spinning drill bit on the end instead of the hot end and leave the motors in place to control the movement of the spinning bit, you've got the basic idea of a CNC machine.

The milling bit removes material instead of adding material. Place a small piece of plywood on the worktable, and a CNC machine (using the familiar g-code) can be instructed to move the milling bit to and fro over the plywood. The Z axis controls how deep into the wood the milling bit cuts. Using a combination of movement along all three axes, you may be able to start picturing how a CNC machine can remove wood (in small amounts), leaving behind an object.

On the simple side, a CNC machine can be used to engrave your name, for example, into a variety of materials—plastic, wood, and metal being the most popular. (Note that you must select the proper milling bit for the material you want to work on.)

On the more advanced side, a milling bit can be used to create intricate designs and even cut out objects from a larger piece of material.

Take a look at Figure 11.1 and you'll see one example of a CNC machine. This is the ShopBot Buddy from ShopBot, a manufacturer of CNC machines.

FIGURE 11.1 This CNC machine uses a milling bit and moves along three axes.

The first thing you'll notice is that it's larger than the Simple 3D printer. Most CNC machines are larger, but some desktop versions (often called milling machines) are just as tiny as the smaller 3D printers.

Most CNC machines work like 3D printers—a series of motors move the milling bit around on the workspace. Unlike the Simple, however, where it's the print bed that moves left and right (instead of the hot end moving left and right), most CNC machines don't move the workspace. Instead, the milling bit moves in all three directions. The milling bit shown in Figure 11.2 moves left and right, forward and backward, and up and down; the workspace stays put.

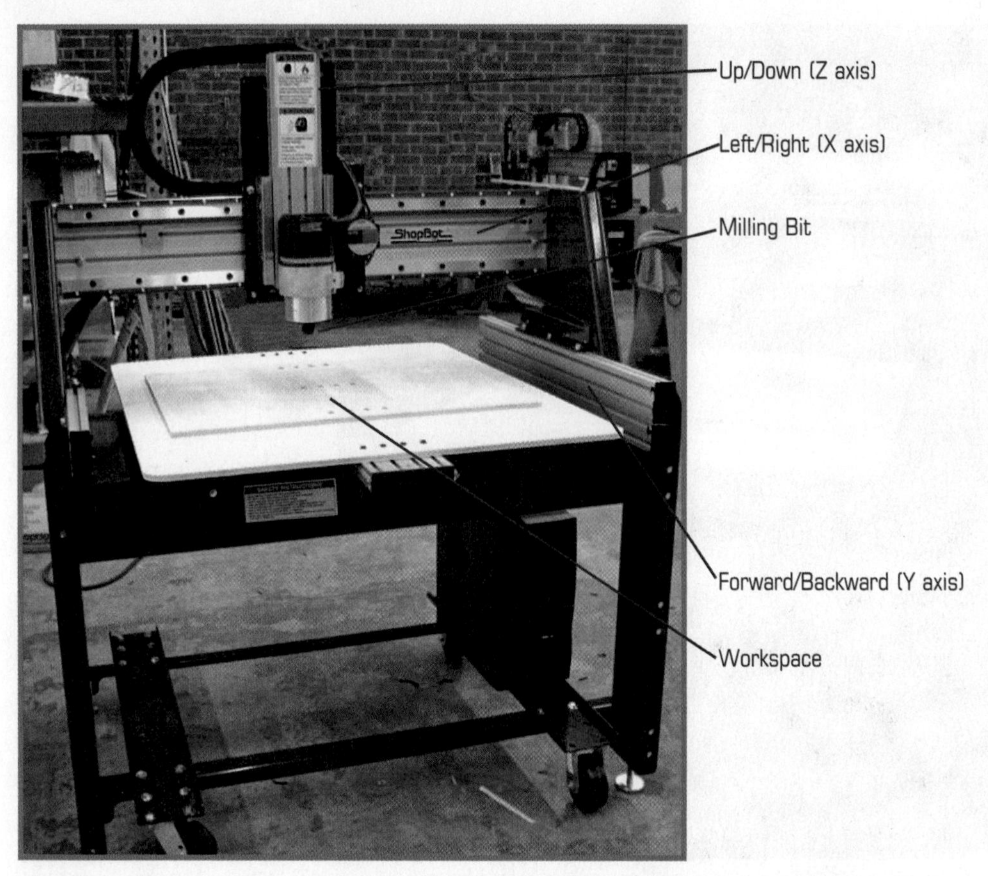

FIGURE 11.2 Movement of the milling bit occurs in all three directions.

This is a 3-axis CNC machine. It moves along the X, Y, and Z axes. But there are more advanced CNC machines called 4-axis that also allow for rotation like a lathe. It's a bit beyond the purposes of this discussion; just be aware that CNC machines have been around longer and are much more advanced, offering more features and tools than 3D printers do.

You're already familiar with how to control a 3D printer with g-code. A CNC machine works the same way, but differences exist. When it comes to controlling a CNC machine with a computer, you're also dealing with depth of cut. 3D printers build up in layers, but CNC machines cut down into material, so the software needed to do this is often more advanced than the Repetier software you learned about in Chapters 5, "First Print with the Simple," and 7, "Creating a 3D Model with Tinkercad."

There's a lot more to know about using a CNC machine than I can provide here, but you'll be happy to know that dozens of books and websites are dedicated to CNC technology. If you're truly interested in expanding your skills, you shouldn't find jumping from 3D printing to milling with a CNC machine too difficult. But just as with 3D printing, you've got to start simple with a CNC machine and continue learning and experimenting to improve your skills with this powerful tool.

NOTE

You won't find too many CNC machines in the price range of low-end 3D printers, and it's not uncommon to spend $2,000 to $10,000 on a hobbyist-level CNC machine. Fortunately, there's an entire hobbyist market out there dedicated to building homemade CNC machines. A good friend of mine, Patrick Hood Daniel, offers a variety of kits and plans if you'd like to try your hand at saving some serious cash. Point a web browser to http://buildyourcnc.com for more information.

Laser Cutter

Lasers. I love that word. It just sounds high-tech.

Imagine for a moment replacing the milling bit in a CNC machine or the hot end in a 3D printer with a powerful laser that can cut through wood, plastic, and other materials. Using a similar method of controlling the motors, you could focus that laser beam onto the workspace and let it cut out shapes and letter curves in whatever material you desire.

It may sound like science fiction, but it's a reality and it's called a laser cutter. Take a look at Figure 11.3 and you'll see one of the more popular laser cutters from Epilog, called the Zing.

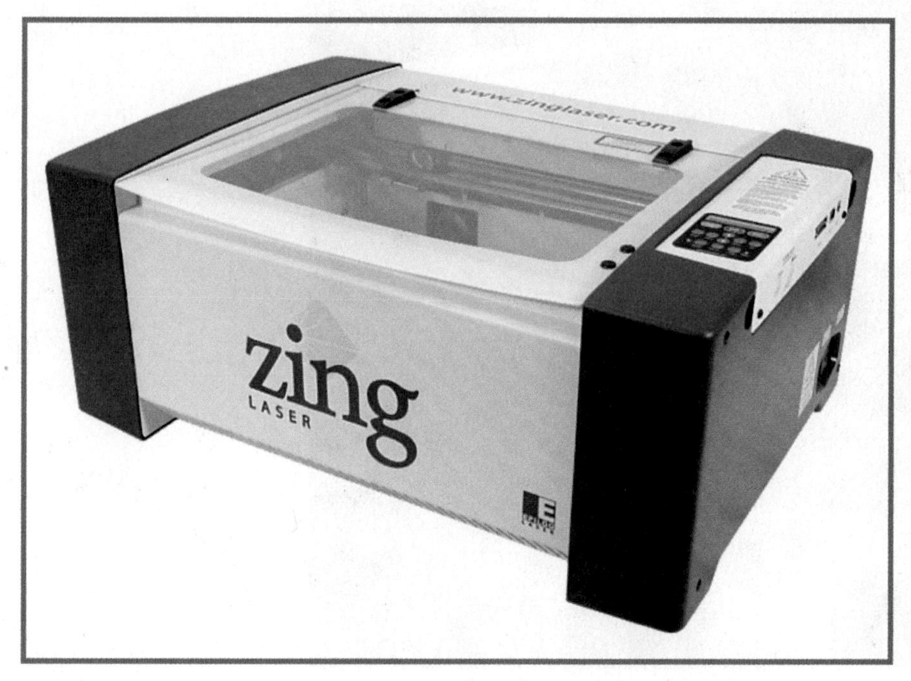

FIGURE 11.3 A laser cutter uses a laser to remove material.

One of the first things you might notice about a laser cutter is how much it looks like a large box. There are a number of reasons for this. First, it is for safety—you don't want someone putting their hands anywhere around a working laser. Second, a lot of materials produce fumes when cut by a high-temperature laser, and the enclosure helps to redirect those fumes using a special ventilation tool that sucks out the fumes and expels them elsewhere (usually outside). Finally, some materials (like glass) can reflect a laser, so the enclosure helps protect eyeballs from an accidental reflection of the beam.

Inside the enclosure you'll find a setup similar to a CNC machine. The workspace stays put, and the laser moves left and right and forward and backward. The only difference is that a laser doesn't need to move up and down—it either cuts all the way through a material, or the software controlling the beam uses a series of pulses that cut only a fixed depth into the material.

As you can imagine, laser cutters are not cheap. Between desktop versions of a CNC machine, a 3D printer, and a laser cutter, you'll definitely spend the most on the laser cutter. It's a technology that hasn't dropped in price as fast as the other two technologies. But it is dropping. Ten years ago a laser cutter would have cost you well over $100,000, but today you can find hobbyist versions for under $10,000. There are even "build your own" versions, such as the one from buildyourcnc.com shown in Figure 11.4, that you can grab for less than $2,000.

FIGURE 11.4 This laser cutter was made from a kit.

Laser cutter special software is also needed to control laser cutters, so you'll typically get this software when you purchase the cutter. You should also be aware that the laser tube that comes inside a laser cutter must eventually be replaced; it's a consumable item, meaning that it has only so many hours of usage. Be certain to inquire not only about the laser tube's expected life before making any purchase of a laser cutter but also about the replacement tube cost.

Plasma Cutter

The last item I want to recommend is one I do not own. I have a CNC machine, a laser cutter, and a few 3D printers, but the plasma cutter remains one of those tools that I'm okay with renting when I need it rather than purchasing. Let me explain.

A handheld plasma cutter is your go-to tool for when you want to cut sheet metal. It's a dangerous tool, and there's definitely a learning curve that must be overcome with some training. But if you're into cutting metal with extremely accurate cuts, this is the tool you want.

Now, take the handheld plasma cutter and think about mounting it in a machine that will let the cutter be directed along the X and Y axes, and you'll have the basic idea of a computer-controlled plasma cutter. Software directs the movement of the plasma cutter, allowing for finer and much more accurate cutting of sheet metal.

In Figure 11.5, you'll see one of the most popular computer-controlled plasma cutter frames on the market—the PlasmaCam. (Point a web browser to www.plasmacam.com and watch a video of the tool in action; it'll blow you away.)

FIGURE 11.5 The PlasmaCam cuts through metal at high speed and with accuracy.

Why do I call it a plasma cutter frame? Because the PlasmaCam is simply a table/workspace and motors that are used to move a handheld plasma cutter that is mounted in the frame. You must purchase a plasma cutter to be mounted in the frame. (There are a variety of plasma cutter manufacturers, so you're not locked into one particular brand.)

Computer-controlled plasma cutters like the PlasmaCam typically come with special software for cutting out two-dimensional objects in sheet metal. Again, if you understand how 3D printers work in terms of moving the working end (hot end) around the workspace, it's not a big jump to understand how a plasma cutter could be directed to cut out objects from metal.

One final caveat: Plasma cutters produce a lot of sparks as they cut metal, and that's one of the reasons why I don't own one—the potential to start fires. Plasma cutters are ideally used in areas away from flammable materials, making them a bit more difficult for the home hobbyist to use safely. When I need some sheet metal cut, I've got a local machine shop that has a variety of tools, including a computer-controlled plasma cutter. I provide them with the file containing the shape I want to have cut out, and they provide the sheet metal and the tool and the final cutout. You may not ever own a plasma cutter, but you should know that there are shops that will rent you the time on one or do the work for you (at a cost, of course).

The Workshop of the Future

I hope you are beginning to get an idea of what someone with the right skills and tools could create. With a 3D printer, a plasma cutter, a laser cutter, and a CNC machine, you'd have quite a collection of advanced tools to make just about anything your mind could come up with.

But if you've got none of these, the 3D printer is definitely the place to start. You'll learn some CAD, spend less money, and have access to a tool that's a bit safer than the other three. If you find that you're enjoying designing and creating things with your 3D printer, you can always expand your horizons and add a new tool (or two, or three) and open up even more opportunities to explore.

Have fun!

3D Printer and Modeling Resources

The information in this appendix is by no means complete. Thousands of websites are dedicated to 3D printing, along with hundreds of books and magazines and enough software companies to fill a dozen pages. What I've done here is pull out some of the ones that were most useful to me during the writing of this book, as well as resources that I believe will help point you in many more directions and expand your own research.

Websites

- Official Printrbot Forum—Printrbot has its own Help site where users can post questions and get help. If you've got a question about your Printrbot, this is definitely the place to start: http://help.printrbot.com/.
- Unofficial Printrbottalk.com Forum—The Printrbottalk.com Forum started when Printrbot was raising funds via its Kickstarter program, and it has continued to be a great place for fans of Printrbot to gather and post photos, ask questions, and more. There's a dedicated area just for the Printrbot Simple, too: http://bit.ly/1c2XuyV.
- Fine-Tuning Your Printrbot Simple—A great forum discussion from printrbottalk.com that offers some additional configuration for the Repetier software, some slightly different than the official Printrbot documentation suggests: http://bit.ly/15J7Gb7.
- Additional 3D Forums—Here are links to some additional 3D forums:
    - Soliforum—www.soliforum.com
    - 3Dprinting-Forums—www.3dprinting-forums.com
    - Make magazine 3DP Forum —http://bit.ly/1bNvXmE
    - RepRap Forum—http://forums.reprap.org
- Shapeways—One of the larger online 3D printing services that will ship your printed model to you. You can even set up a shop and sell items you've designed: www.shapeways.com.
- Sculpteo—Another great online 3D printing service that will take your uploaded model, print it, and ship it to you. Also supports selling your creations: www.sculpteo.com.
- Ponoko—Here's one more excellent online 3D printing service for you to investigate. Prints and ships your uploaded models and has a seller's area as well: https://www.ponoko.com.

- Thingiverse—Quite possibly the largest collection of digital files suitable for 3D printing, Thingiverse can keep you busy browsing for hours. Download and print items you find, and upload your own: www.thingiverse.com.
- 3Ders.org—Get prices on 3D printers, read articles, and stay up to date with the latest 3D printing news. Includes some great videos of different types of 3D printing technology: http://www.3ders.org.

Books

- *3D CAD with Autodesk 123D,* by Jesse Harrington Au—Covers the entire 123D family of software. ISBN: 1449343015.
- *Getting Started with RepRap,* by Josef Prusa—A beginner's guide to the DIY RepRap 3D printer. ISBN: 1457182963.
- *Design & Modeling for 3D Printing,* by Matthew Griffin— Discussion and activities to improve your modeling skills. ISBN: 1449359175.
- *Make: Ultimate Guide to 3D Printing*—An annually updated collection of 3D printer reviews, how-to articles, and more. ISBN: 1449357377 (2012 edition).

Software

Below you'll find links to the developers of applications that can be used to create 3D models to print with your 3D printer. I'm not including prices because they change, but be aware that many of the applications are not free.

- Autodesk—www.autodesk.com*
- Blender—www.blender.org
- Solidworks—www.solidworks.com
- GrabCad—http://grabcad.com
- Rhino3D—www.rhino3d.com
- 3Dtin—www.3dtin.com/
- FreeCad—http://sourceforge.net/projects/free-cad/
- LibreCad—http://librecad.org/cms/home.html
- SketchUp—www.sketchup.com
- Art of Illusion—www.artofillusion.org
- Sculptris—http://pixologic.com/sculptris/
- Wings3D—www.wings3d.com
- OpenSCAD— http://www.openscad.org

* The AutoDesk software is free to students and educators—visit http://students.autodesk.com for more details.

3D Printers

The number of 3D printing companies seems to be growing almost daily. It's impossible to list all the sellers and their most current prices, but I have found that the 3ders.org website does a great job of keeping up (as best it can) with the latest 3D printers and their prices. Point your web browser to the following address for a long list of sellers and prices:

http://www.3ders.org/pricecompare/3dprinters/

3D Printing Supplies

- Amazon.com—yes, Amazon.com—has an entire 3D printer supply category that includes not just printers but also replacement parts, filament, and more, which is especially useful if you're considering building your own 3DP: http://amzn.to/11rnXOX.
- Ultimachine—They sell printers, but they're also a good one-stop shop for supplies: https://ultimachine.com.
- eBay—Famous for online auctions, eBay also offers 3D printing sellers a place to offer their wares: http://bit.ly/12jenUn.

Index

Symbols

A

B

C

D

E

F

G

H

I-J

K

L

M

N

O

P

Q

R

S

T

U-V

W

X

Y

Z